HIGH
TEMPERATURE
COATINGS

This book is to be returned on or before
the last date stamped below.

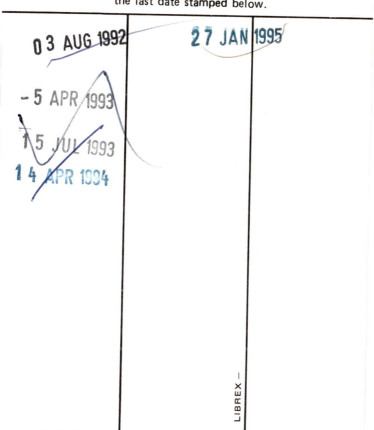

HIGH
TEMPERATURE
COATINGS

Proceedings of a symposium on High Temperature Coatings,
held during the 1986 Fall Meeting of The Metallurgical Society in
Orlando, Florida, October 7-9, 1986.

Edited by

M. Khobaib
University of Dayton
Dayton, Ohio

R.C. Krutenat
AVCO Specialty Materials, TEXTRON
Lowell, Massachusetts

A Publication of 🏛️ The Metallurgical Society, Inc.

A Publication of The Metallurgical Society, Inc.
420 Commonwealth Drive
Warrendale, Pennsylvania 15086
(412) 776-9000

Printed in the United States of America.
Library of Congress Catalogue Number 87-20438
ISBN NUMBER 0-87339-068-7

© 1987

Foreword

This book covers a broad spectrum of topics covering the subject of high temperature coatings. It incorporates 14 papers presented at the symposium, "High Temperature Coatings," held in Orlando, Florida, October 7-9, 1986, during the Fall Meeting of The Metallurgical Society of AIME. The symposium addressed current research trends, application opportunities for high temperature aluminide coatings, and the mechanistic aspects of coating/substrate interaction. The symposium brought together researchers from all over the United States which resulted in fruitful interaction through intelligent discussions of the important aspects of high temperature coatings.

A majority of the papers deal with coatings related to turbine engine applications. This is an area of great interest to aircraft engine manufacturers who continually desire new, improved, and reliable coatings for high performance at elevated temperatures. The remaining papers deal with methods for applying coatings and the performance and mechanical behavior of a variety of coatings.

The application and performance of thermal barrier coating is discussed by a select group of authors. Coating/substrate interactions and the interfacial stability is also extensively discussed in terms of interdiffusion, microstructural stability, formation of various phases, etc. A novel idea for inspection of integrity of a coating bond is also covered along with the topic of use of chromized coatings for protection against stress corrosion cracking in space shuttle applications.

The editors extend their appreciation to Dr. Russel H. Jones, Battelle Pacific Northwest Laboratories who, as chairman of the Environmental Effects Committee, gave his approval to plan, organize, and publish the proceedings of the symposium.

M. Khobaib
University of Dayton
Research Institute
Dayton, Ohio 45469

R.C. Krutenat
AVCO Specialty Materials, TEXTRON
2 Industrial Avenue
Lowell, Massachusetts 01851

Table of Contents

COATABILITY OF HIGH GAMMA-PRIME

ODS SUPERALLOYS

R. C. Benn
Inco Alloys International, Inc.
Huntington Alloys
P. O. Box 1958
Huntington, West Virginia   25720

P. Deb and D. H. Boone
Department of Mechanical Engineering
Naval Postgraduate School
Monterey, California   93943-5000

Abstract

ODS alloy substrates with varying compositions were coated with two types of coatings (1) an outward type (two zone) vapor phase aluminide and (2) an advanced NiCoCrAlY overlay. These coated specimens were exposed to a cyclic oxidation environment at 1100°C for times up to 415 hours (i.e., 500 cycles). The diffusion aluminide specimens were tested for 146 hours because of excessive porosity formation at that point. Additional uncoated and coated specimens were hot corrosion tested in a Burner Rig at 927°C for 300 hours (uncoated specimens removed after 168 hours). For the diffusion aluminide coatings, the presence of higher levels of aluminum (and to a lesser extent yttria) in the substrate alloy significantly reduced the formation of porosity, both in size and distribution in the inner coating zone. Both overlay coatings exhibited significantly better resistance to porosity formation as compared to the diffusion aluminide coating. Plasma sprayed coatings exhibited porosity in the outer as well as in the inner coating zones while electron beam physical vapor deposited coatings exhibited porosity only in the inner coating zone. Like diffusion aluminide coating phenomena, the presence of higher levels of aluminum in the substrate significantly reduced the formation of coating associated porosity. The combined effect of increased levels of chromium and tantalum was found to be rather complex and generally beneficial in the aluminides as well as the overlays. There was no beneficial effect of cobalt additions to the substrate alloy on porosity formation. The "layered" structure of the plasma sprayed overlay coating appeared to disperse vacancy migration and thereby reduce the formation of any gross porosity. Link-up of pits in the coating surface with porosity at the substrate-coating interface lead to substrate hot-corrosion attack on some aluminide coated alloys. The hot corrosion resistance of the overlay coated alloys was significantly better with no substrate attack. Further tailoring of coating and substrate composition together with the use of diffusion limitors suggests the potential for porosity-free coated ODS components.

1

Introduction

Mechanical alloying (1) has yielded a new class of superalloys combining gamma-prime precipitation hardening for intermediate temperature strength with oxide dispersion strengthening (ODS) for high temperature strength (2-8). While the oxidation and hot corrosion resistance of these alloys are equal to or better than conventional superalloys of similar composition (1,6), protective coatings remain desirable for extended service life or for service in particularly harsh environments. In the present investigation, several coating systems were applied to a series of high strength ODS-Mechanically Alloyed alloys having 50 to 80% gamma-prime by volume (3-8). Such alloys are known to be more compatible with coatings than ODS alloys with lower levels of Al (9).

The availability of suitable coatings required to provide protection for a sufficient length of time to match the alloys high temperature strength capabilities has been in question. This coatability question can arise because of the selective nature of the diffusing species in the intermetallic coating phases, sometimes resulting in the formation of porosity which in turn can result in premature loss of the coating and subsequent substrate attack (1,10). The effect of substrate composition and processing on this interaction is not well understood and has not been studied.

In the present investigation, an effort has been made to assess the effect of ODS-MA substrate composition on the coating diffusional stability and hot corrosion resistance of two archetype coatings, (i) an outward type diffusion aluminide and (ii) an advanced overlay composition.

Experimental Technique

Two types of coatings (i) an outward (two zone) vapor phase aluminide and (ii) an advanced NiCoCrAlY overlay were applied to a series of ODS-MA alloys, whose chemical compositions are listed in Table I. These samples were exposed to cyclic oxidation at 1100°C for times up to 416 hours (500 cycles). The outward aluminide coating was selected on the basis of

Table I. Chemical Compositions of ODS Alloys

Alloy	Cr	Al	W	Mo	Ta	Nb	Co	Hf	Ti	B	Zr	Yttria
51	9.3	8.5	6.6	3.4	-	-	-	-	-	0.01	0.15	1.1
69	9.3	7.0	8.0	2.0	1.0	-	-	-	-	0.01	0.15	1.1
70	11.0	7.0	6.5	1.7	1.6	-	-	-	-	0.01	0.15	1.1
71	9.3	8.5	8.0	2.0	1.0	-	-	-	-	0.01	0.15	1.1
72	11.0	8.5	6.5	1.7	1.6	-	-	-	-	0.01	0.15	1.1
74	9.3	8.5	6.6	3.4	-	-	5	-	-	0.01	0.15	1.1
76	9.3	8.5	6.6	3.4	-	-	-	-	-	0.01	0.15	0.6
77	12.0	7.3	6.0	3.0	2.0	1.0	-	0.5	-	0.01	0.15	1.1
MA 6000	15.0	4.5	4.0	2.0	2.0	-	-	-	2.5	0.01	0.15	1.1
78	15.0	7.0	4.0	2.0	2.0	-	-	-	-	0.01	0.15	1.1

1. Effect of Yttria: Alloy 76 vs. 51
2. Effect of Al: Alloy 71 vs. 69
                        Alloy 72 vs. 70
                        Alloy 78 vs. MA 6000 [:Effect of Ti]
3. Effect of Cr + Ta: Alloy 70 vs. 69
                             Alloy 72 vs. 71
4. Effect of Co: Alloy 74 vs. 51

earlier work (9) relating ODS alloy composition to aluminide coatability. In that study, it was found that in addition to the strong beneficial effect found for aluminum additions to the substrate, the outward two zone type aluminide was better than the inward type coating. A standard commercially used NiCoCrAlY overlay coating was applied by (i) electron beam physical vapor deposition (EB-PVD) and (ii) vacuum plasma spray (VPS) techniques followed by standard post-coating peening and heat treatment.

Some selected substrates coated with either the aluminide or the overlays were also exposed to isothermal testing at 1100°C for times up to 250 hours. Both tests (cyclic and isothermal oxidation) were performed in furnaces where temperatures were controlled to ±5°C. In the cyclic test, the specimens were heated for 50 minutes at 1100°C and then cooled 10 minutes outside the furnace prior to reheating. Wherever possible duplicate tests on each substrate were performed.

Diffusion aluminide coatings were removed from the test after a maximum of 146 hours (175 cycles) while the two overlays were tested for a total of 416 hours (500 cycles). Although examinations were carried out at a number of time intervals, the results presented in this paper are based primarily on examination after 146 and 416 hours exposure. In an attempt to quantify the level of porosity observed, a semi-quantitative procedure was employed. The average linear density of porosity was estimated by a point counting technique while the average size of the porosity was ranked into three arbitrary categories, (i) fine, e.g. $\lesssim$5µm; (ii) medium, e.g. ~10µm; and (iii) coarse, e.g. $\gtrsim$15um by comparing their relative size at constant magnification.

Burner Rig sulfidation tests were conducted at 927°C for 168 h (uncoated specimens) to 300 h (coated specimens). The rig used corresponded to the G.E. Lynn low velocity burner rig (11). A 30:1 air/fuel (JP-5, 0.3% S, 5 ppm standard sea water) ratio was used. The specimens had a 2 min. cycle per hour (i.e., 58 min. in the flame, 2 min. out in air).

Results and Discussion

Selection of Alloys

Alloys were selected from among a number of gamma-prime containing ODS-MA alloys in order to determine the effects of base alloy composition on the coatings' compatibility and surface stability. Specifically, the alloys varied in yttria content, aluminum content, chromium + tantalum content, titanium content, and cobalt content. Yttria and aluminum were expected to strongly affect both coatability and surface stability, while the other components were likely to affect surface stability only.

Alloy 51 is an alloy that has been extensively characterized (6) and served as the baseline alloy in this program. Of particular interest was its relatively good coatability as suggested in a separate study (9). The high aluminum content of alloy 51 (8.5%) apparently reduced interdiffusion and void formation, which leads to spallation. This alloy received the full range of coatings.

Alloy 69 differs from alloy 51 most significantly in having a lower aluminum level (7.0%). It also contains 1.0% tantalum and has somewhat different tungsten and molybdenum levels. Alloy 70 contains higher chromium and tantalum than 69 and reduced refractory metal levels. Alloys 71 and 72 are similar to 69 and 70, respectively, but retain the aluminum level of 51 (8.5%). These four alloys were selected to collectively allow the determination of the effects of aluminum and chromium + tantalum.

Alloy 74 is a modification of alloy 51 having a 5% cobalt addition. This element is a common component of conventional superalloys but has not been well characterized in ODS alloys.

Alloy 76, with its lower level of yttria (0.6%) compared to alloy 51 (1.1%; the standard level for these alloys) was chosen to reveal the effects of dispersoid content on coatability and surface stability. The difficulty encountered in coating first generation ODS-MA alloys has been known for some time (10), but the effect of varying levels of dispersoid has not been studied specifically.

INCONEL* alloy MA 6000 is an ODS superalloy with a moderate level of gamma-prime that is presently in commercial production. It features the lowest Al level of the selected alloys and is also the only alloy with Ti. Alloy 78 is the same as INCONEL alloy MA 6000 but with a complete substitution of additional Al for Ti, giving it the same Al level as alloys 69 and 70 (7%). The previous studies showed that this level of Al may provide slightly greater strength than the 8.5% of alloy 51. The same has been shown for alloy 78, which also promises to have significantly better hot corrosion resistance with its high Cr level. INCONEL alloy MA 6000 and alloy 78 should provide interesting data on the combined effect of Al and Ti.

Cyclic Oxidation

Diffusion Aluminide Coatings

Previous exposures of aluminide coated ODS/MA alloys reported porosity in the coating at times as short as 6 hours at 1180°C for some systems (Al free) in the as-coated and diffused condition (9). It was noted the porosity, when present, occurs in the so-called inner diffusion

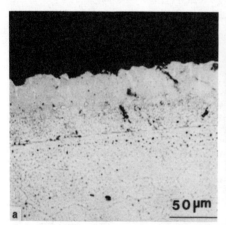

Figure 1. Micrographs of diffusion aluminide coating on INCONEL alloy MA 6000 substrate: (a) 80h exposure at 1100°C and (b) 146h exposure at 1100°C.

---

*INCONEL is a trademark of the Inco family of companies.

4

zone, a region of the coating comprised of the substrate with Ni (Co) extracted by the aluminum. No porosity was observed in the outer coating zone which is Y_2O_3 free. In this investigation selected samples were examined at times as low as 66 hours to assess the presence and development of porosity. Figure 1a shows no porosity after 66 hours exposure. All of these specimens showed significant porosity after 146 hours in the interdiffusion zone shown in Figure 1b. The estimated density of porosity for the aluminide coating is given in Table II.

TABLE II. QUANTITATIVE ESTIMATION OF POROSITY IN DIFFUSION ALUMINIDE COATING AFTER 146 HOURS EXPOSURE AT 1100°C

Specimen (alloy)	Porosity Present in the Inner Coating Zone and Their Relative Size
76	50% porosity, coarse in size
51	10% porosity, fine in size much smaller than alloy 76
71	15% porosity, medium in size
69	70% porosity, coarse in size
72	60% porosity, coarse in size
70	90% porosity, coarse in size
78	15% porosity, medium in size
MA 6000	100% porosity, coarse in size
70	90% porosity, coarse in size
69	70% porosity, coarse in size
72	60% porosity, coarse in size
71	15% porosity, medium in size
74	15% porosity, coarse in size
51	10% porosity, fine in size much smaller than alloy 74

1. The Effect of Y_2O_3

It was found that higher substrate levels of yttria, such as 1.1% (i.e., alloy 51) have an apparent beneficial effect in reducing porosity formation over the lower yttria (0.6%, i.e., alloy 76). It was also seen that the average size of the porosity in alloy 51 was much smaller than that found in alloy 76 as shown in Figure 2. Figures 2(a) and 2(b) exhibit porosity near the interdiffusion zone. This apparent benefit is surprising since Y_2O_3 free substrates are found to be porosity free. The addition of Y_2O_3 and other dispersions (i.e., ThO_2) tends to accelerate the formation of coating induced porosity. The possibility exists that increased dispersant levels may not decrease the amount of porosity but may act to nucleate its precipitation over a broader range on a finer scale and thus it appears less severe.

2. The Effect of Aluminum

The higher level of substrate aluminum content has a significant contribution in preventing the formation of porosity in the so-called inner diffusion zone of the coating. This Al effect was previously reported and related to the effect of the higher Al in producing more β(NiAl) phase for a given amount of Ni (and Co) outward diffusion through

5

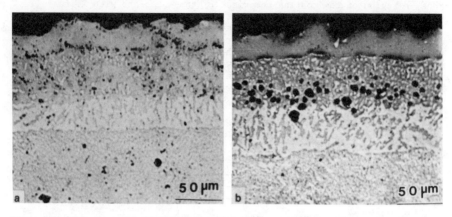

Figure 2. Micrographs of diffusion aluminide coating on alloys 51 and 76 after 146h exposure at 1100°C: (a) fine porosity in alloy 51 and (b) coarse porosity in alloy 76.

the hypostoichiometric β (NiAl) phase. Simultaneous precipitation of the other strengthening elements with low solubilities in the β phase may also play a role. The Ni (and Co) depleted substrate will transform to β (NiAl) with the subsequent precipitation of refractory metal-rich phases and a volume change (expansion). This volume change has been proposed to compensate, at least in part, for the vacancy flux and the resulting porosity formation in this zone.

Of the six alloys which are modifications of alloy 51 (i.e., alloy 71 vs 69, 72 vs 70 and 78 vs INCONEL alloy MA 6000), alloys 71 and 78 had about 15% porosity in the inner coating zone and the size of the porosity was finer than that in alloys 69, 70, and INCONEL alloy MA 6000 respectively which had 70, 90, and 100% porosity.

3. The Effect of Cr and Ta

Alloys 69 and 72 with 9.3% Cr exhibited a better resistance to the formation of porosity than alloys 70 and 72 with 11% Cr. It was found that alloy 71 exhibited about 15% porosity after 145 hours exposure while alloy 69 had about 70% porosity.

4. The Effect of Co

It appears that there is no beneficial effect of adding Co to the substrate to prevent forming porosity. Alloy 74 (with 5% Co) exhibited about 15% porosity (coarse in size) as compared to cobalt-free alloy 51 which exhibited about 10% porosity (average size much smaller than that of alloy 74). INCONEL alloy MA 6000 is also Co-free. Cobalt would be expected to diffuse with Ni outward through the β (NiAl) phase although the driving energy for Co-Al interaction is less than for Ni-Al. The other possible effect of Co is on the Al content of the $\gamma' - \beta$ phase boundary. As previously noted the promotion of increased β phase and concomitant refractory metal-rich phase precipitation with Ni (and Co) extraction is predicted to reduce the porosity formation. Precipitate size and distribution will also affect porosity size.

Summarizing the effect of substrate composition on aluminide coating stability, it can be concluded that the alloys 51, 71 and 78 have better resistance to porosity formation than other substrates. Of these three alloys, alloy 51 seems to have the optimum composition. This is the only system where one can find the combination of all alloying elements which prevent porosity formation.

This observation is not too surprising if the postulation is correct that the beneficial effects of Al are to reduce the amount of Ni extraction required to precipitate β and other phases. Alloy 51 was formulated for a high volume fraction of γ' phase, and therefore its composition lies near the β phase boundary. A given amount of nickel removal would produce a larger amount of intermetallic phase precipitation in this alloy than in the other, lower Al and γ' phase content alloys.

Overlay Coatings

The overlay coatings were originally developed to be substrate independent, and therefore their use for the protection of the ODS/MA alloys would appear ideal. If no interdiffusion occurs, the formation of porosity, at least Kirkendall type, would not be expected. However, in an effort to optimize specific properties of the overlay coatings such as hot corrosion resistance, high temperature oxidation resistance and/or mechanical properties, a number of overlay coating compositions have been formulated. Some of these have been studied on the ODS/MA alloys with reported improved diffusional stability (12-13) when compared to the aluminides.

As mentioned earlier, the overlay coatings exhibited no detectable porosity after 250 hours (300 cycles) exposure (see Figures 3 and 4) while diffusion aluminide coatings showed porosity at 66 hours and excessive porosity after only 146 hours exposure at the same temperature. It is significant that the porosity was observed in the outer zone of the overlay coating applied by the VPS technique after 340 hours exposure while the EB-PVD coating did not show any porosity in the outer zone (see Figures 5 and 6). Note again that when substrate-associated porosity was

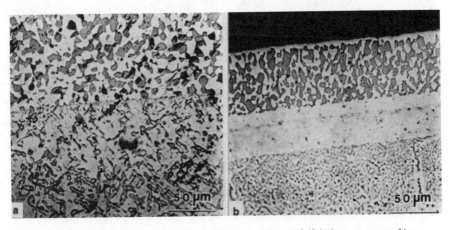

Figure 3. Micrographs of EB-PVD overlay coating exhibiting no porosity after 250h exposure at 1100°C: (a) alloy 77 and (b) alloy 78.

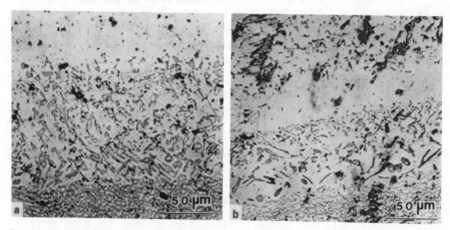

Figure 4. Micrographs of PS overlay coating exhibiting no porosity after 250h exposure at 1100°C: (a) alloy 51 and (b) alloy 78.

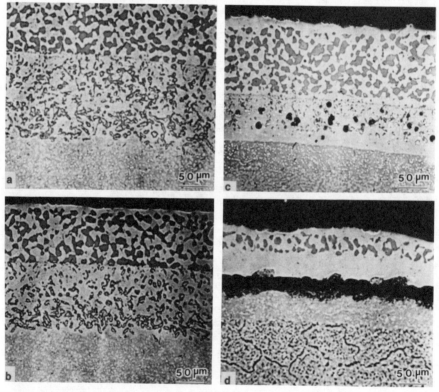

Figure 5. Micrographs of EB-PVD overlay coating exposed 340h at 1100°C exhibiting: (a) no porosity on alloy 72, (b) fine porosity on alloy 70, (c) medium size porosity on alloy 69, and (d) coarse size porosity on INCONEL alloy MA 6000.

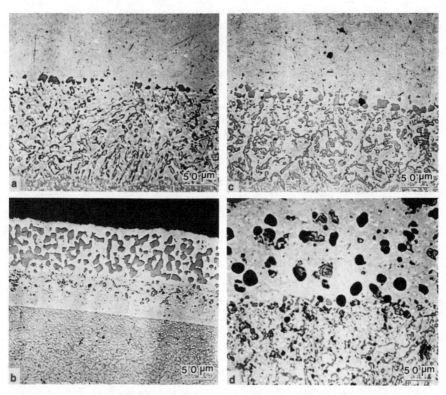

Figure 6. Micrographs of PS overlay coating exposed 340h at 1100°C exhibiting: (a) no porosity on alloy 76, (b) fine size porosity on alloy 71, (c) medium size porosity on alloy 72, and (d) coarse size porosity on alloy 74.

TABLE III. QUANTITATIVE ESTIMATION OF POROSITY IN OVERLAY COATING AFTER 340 HOURS EXPOSURE AT 1100°C

Specimen (alloy)	EB-PVD Porosity Present in the inner coating zone and their relative size.	Plasma Spray Porosity present in the outer and inner coating zone and relative size.
76	About 5% medium size porosity.	No porosity.
51	About 10% medium size porosity.	-
71	No porosity.	10% porosity, fine size.
69	40% porosity, coarse in size.	10% porosity, fine size.
72	No porosity.	15% med. size porosity in inner zone and 20% porosity in outer zone.
70	Less than 5% fine size porosity.	
78	No porosity.	-
MA 6000	100% porosity.	-
70	Less than 5% fine size porosity.	No porosity.
69	40% porosity, coarse in size.	10% porosity, fine size.
72	No porosity.	No porosity.
71	No porosity.	Less than 5% fine porosity.
74	20% medium size porosity.	Less than 10% medium porosity in inner zone and 80% coarse porosity in the outer zone.
51	About 10% medium size porosity.	-

9

found in these systems it occurred in the original ODS-MA substrate
material and not in the overlayed coating. Both techniques produced
better resistance to porosity formation except for substrate alloys 74, 69
and INCONEL alloy MA 6000 (see Table III). Of these three substrates
INCONEL alloy MA 6000 showed the highest porosity level. ODS-MA alloys
51, 70 and 69 coated by both techniques also showed resistance to
porosity formation even after 416 hours exposure (see Table IV) shown in
Figures 7 and 8.

**TABLE IV. QUANTITATIVE ESTIMATION OF POROSITY IN OVERLAY COATING
AFTER 416 HOURS EXPOSURE AT 1100°C**

Specimen (alloy)	EB-PVD Porosity Present in the inner coating zone and their relative size.	Plasma Spray Porosity present in the outer and inner coating zone and relative size.
76	10% coarse porosity	-
51	10% coarse porosity.	10% med. size porosity in the inner zone and 5% coarse porosity in the outer zone.
71	50% medium size porosity.	-
69	50% coarse size porosity.	25% med. size porosity in inner zone and 30% coarse porosity in outer zone.
72	30% medium size porosity.	20% med. size porosity in inner zone and 20% coarse porosity in outer zone.
70	15% medium size porosity.	20% fine size porosity in the inner zone and 15% coarse size porosity in outer zone.
78	70% coarse size porosity.	60% coarse size porosity in outer zone and 20% coarse size porosity in inner zone.
MA 6000	100% porosity.	100% porosity in the inner zone and 80% coarse porosity in the outer zone.
70	15% medium size porosity.	20% fine porosity in inner zone and 15% coarse porosity in outer zone.
69	50% coarse size porosity.	15% fine porosity in the inner zone and 15% coarse porosity in the outer zone.
72	30% medium size porosity.	20% medium size porosity in inner zone and 30% coarse porosity in the outer zone.
71	50% medium size porosity.	-
74	25% medium size porosity.	15% medium size porosity in inner zone and 60% coarse porosity in the outer zone.
51	10% coarse size porosity.	10% medium size porosity in inner zone and 5% coarse porosity in the outer zone.

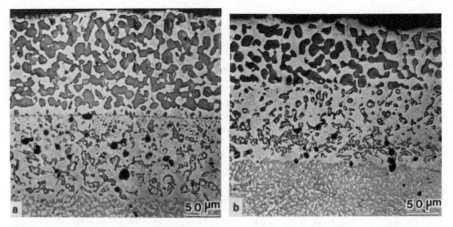

Figure 7. Micrographs of EB-PVD overlay coating exposed 416h at 1100°C exhibiting medium size porosity in the inner zone (1) alloy 51 and (b) alloy 70.

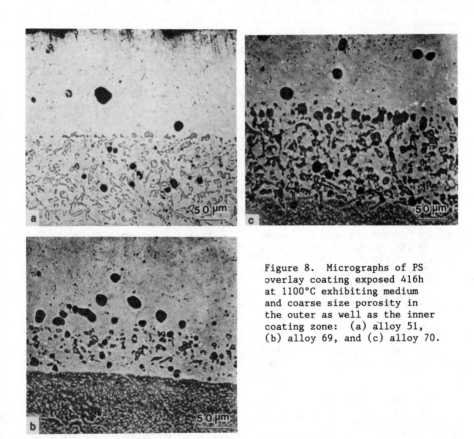

Figure 8. Micrographs of PS overlay coating exposed 416h at 1100°C exhibiting medium and coarse size porosity in the outer as well as the inner coating zone: (a) alloy 51, (b) alloy 69, and (c) alloy 70.

11

The influence of substrate alloying elements such as Al, Y_2O_3, (Ca + Ta) and Co on the diffusional stability of the overlay coating is discussed in greater detail in the following section.

1. The Effect of Y_2O_3 Dispersion

For the overlay coatings, no effect of yttria level was found. The ODS-MA alloys (i.e., 76 and 51) coated by both the processes exhibited essentially identical behavior (see Tables III and IV).

2. The Effect of Aluminum

Comparing both the overlay coating processes, the presence of aluminum produced better resistance to porosity formation even after 416 hours (500 cycles) exposure. ODS-MA alloys containing varying amounts of aluminum (i.e., 69, 70,71, 72, 78 and INCONEL alloy MA 6000) exhibited porosity in the interdiffusion zone. Of these six alloys, alloy 70 seems to be optimum substrate composition for the overlay coating produced by either of the processes. This is consistent with the effect of Al found for the aluminides and suggests a similar porosity formation mechanism. It appears from these results that the presence of higher levels of chromium in conjunction with aluminum is also beneficial in reducing the formation of porosity.

3. The Effect of (Cr + Ta)

The addition of Cr + Ta reduced porosity formation in the inter-diffusion zone of alloys 69 and 71 for VPS and 70 and 72 for EB-PVD overlay coating respectively. It appears that the alloys 69 (VPS coated) and 70 (EB-PVD coated) seem to be promising so far as overlay coating stability is concerned.

4. The Effect of Co

Cobalt level was found to have little influence on porosity formation produced by both the processes.

Summarizing, alloys 51, 69 and 70, as compared to other ODS-MA alloys, have compositions which produced the least amount of porosity when coated with this overlay composition. A higher level of substrate aluminum content reduced the amount of porosity (alloy 51). The combined level of Cr + Ta also reduced porosity (alloy 70) even in conjunction with a lower level of aluminum. Yttria level within the limited range studied had no significant effect as far as average size of the porosity and the amount of porosity that was observed. The presence of Co in the substrate had little effect in reducing porosity as was also observed for the aluminide coatings. The implication of these observations are that coating substrate interdiffusion (primarily extraction of Ni) is the cause of porosity, and substrate (and coating) variables that reduce this diffusional flux reduce or eliminate the porosity. It can be noted that porosity formation in and associated with the protective coatings is not limited to ODS/MA substrates but is seen in other systems such as the aluminized aluminum-free cobalt superalloys as well.

A further verification of these conclusions was the observation of porosity in the Ni depleted substrate zone below the β (NiAl)-containing interdiffusion zone. In this zone, sufficient Ni was withdrawn to allow β (NiAl) to precipitate with refractory metals-rich phases. However, below this zone, while Ni was removed, the Al level did not reach a critical value

TABLE IV. QUANTITATIVE ESTIMATION OF POROSITY IN OVERLAY COATING AFTER 416 HOURS EXPOSURE AT 1100°C

Specimen (alloy)	EB-PVD Porosity Present in the inner coating zone and their relative size.	Plasma Spray Porosity present in the outer and inner coating zone and relative size.
76	10% coarse porosity	-
51	10% coarse porosity.	10% med. size porosity in the inner zone and 5% coarse porosity in the outer zone.
71	50% medium size porosity.	-
69	50% coarse size porosity.	25% med. size porosity in inner zone and 30% coarse porosity in outer zone.
72	30% medium size porosity.	20% med. size porosity in inner zone and 20% coarse porosity in outer zone.
70	15% medium size porosity.	20% fine size porosity in the inner zone and 15% coarse size porosity in outer zone.
78	70% coarse size porosity.	60% coarse size porosity in outer zone and 20% coarse size porosity in inner zone.
MA 6000	100% porosity.	100% porosity in the inner zone and 80% coarse porosity in the outer zone.
70	15% medium size porosity.	20% fine porosity in inner zone and 15% coarse porosity in outer zone.
69	50% coarse size porosity.	15% fine porosity in the inner zone and 15% coarse porosity in the outer zone.
72	30% medium size porosity.	20% medium size porosity in inner zone and 30% coarse porosity in the outer zone.
71	50% medium size porosity.	-
74	25% medium size porosity.	15% medium size porosity in inner zone and 60% coarse porosity in the outer zone.
51	10% coarse size porosity.	10% medium size porosity in inner zone and 5% coarse porosity in the outer zone.

to form β (NiAl). Therefore, the possibility of greater porosity formation in this zone exists because of the absence of the volume compensating effects.

When comparing levels of porosity produced by the overlays, minor differences were found between the VPS and EB-PVD processes. The "layered" structure of the VPS coating appeared to offer a grain boundary-type sink for vacancy diffusion and thus reduce the formation of gross porosity. However, extensive porosity was observed in the outer zone of the VPS overlay coating. No porosity was observed in the outer coating zone of the EB-PVD overlay coating.

In addition, the formation of the β (NiAl) phase and refractory metal-rich inner zone appeared to act in some manner as a diffusion barrier as well as serve to compensate for the Kirkendahl induced porosity forming in this zone.

Substrate elements beneficial in reducing porosity are effective for both the overlay and aluminide coatings although more diffusion and hence porosity is found for the same substrates with the aluminides. It should be noted that the occurrence of porosity in coated ODS-MA alloys has been somewhat variable with some reports of porosity-free aluminide test results (14). In an attempt to determine whether this was related to test conditions, an outward aluminide coating on INCONEL alloy MA 6000 superalloy substrate was tested under isothermal as well as cyclic oxidation at 1100°C for times up to 210 hours. The results, described in detail elsewhere (15), appear to indicate that there was no significant effect of cyclic exposure in accelerating porosity formation.

Burner Rig Corrosion Studies

The test samples were bedded into the specimen carousel using refractory cement as shown in the schematic of the Burner Rig assembly (Figure 9). Hot corrosion attack on exposed Burner Rig specimens was measured via metallographic examination of transverse sections according to Figure 10. Details of the hot corrosion attack are given in Table V.

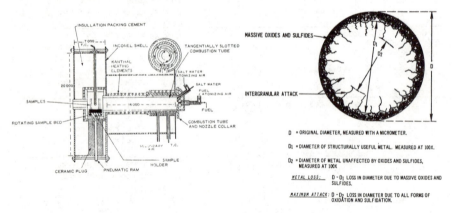

Figure 9.   Burner Rig schematic.

Figure 10.   Method of measuring hot corrosion attack.

In general many of the structural changes that occurred during oxidation exposure were also observed in the exposed Burner Rig specimens but to a lesser degree commensurate with the lower exposure temperature (927°C vs 1100°C). The trend of improved hot corrosion resistance in the uncoated specimens with, inter alia, increasing chromium (and tantalum) content was observed ranging from, say, alloy 51 (9.3% Cr) through INCONEL alloy MA 6000 (15% Cr + 2% Ta) to IN 939 (22.5% Cr + 1.4% Ta). The

TABLE V. BURNER RIG HOT CORROSION RESULTS

	Uncoated (1)		Coated PWA 275 Outward Aluminide (1)		KB-PVD NiCoCrAlY (2)		LPPS NiCoCrAlY (2)	
Alloy	Metal Loss (mm)	Max. Attack (mm)	Metal Loss (mm)	Max. Attack (mm)	Metal Loss** (mm)	Max. Attack** (mm)	Metal Loss (mm)	Max. Attack (mm)
51	D	D	0.020	0.025*	0.002	0.015	0.043	0.076
69	2.995	2.995	0.015	0.036*	0.018	0.038	0.035	0.056
70	0.378	0.378	0.018	0.042	0.008	0.023	0.038	0.061
71	0.630	0.630	0.023	0.063	0.033	0.046	0.038	0.061
72	0.510	0.607	0.023	0.058	0.025	0.033	0.028	0.048
74	D	D	0.025	0.074*	0.015	0.051	0.056	0.089
76	D	D	0.025	0.068*	0.020	0.048	0.066	0.076
77	0.641	1.171	0.018	0.066	0.010	0.066	NA	NA
78	0.020	0.074	0.025	0.046	0.020	0.056	0.030	0.063
MA 6000 (1)	0.023	0.051	0.010	0.038	0.010	0.099	0.041	0.053
MA 6000 (2)	0.039	0.053						
IN-939 (1)	0.025	0.035						
IN-939 (2)	0.020	0.045						
IN-738 (1)	0.028	0.046						

Coating Thickness (Nominal Start)

Aluminide   ~0.012 mm (0.0005 in.)
EB-PVD NiCoCrAlY ~0.125 mm (0.005 in.)
LPPS NiCoCrAlY   ~0.190 mm (0.0075 in.)

NOTES: (1) Run 1
(2) Run 2
D   Destroyed
*   Penetration through coating to substrate.
** End section attack.
NA = Not available.

corrosion attack on most of the uncoated low % Cr alloys was sufficient at 168 h to warrant removal from the rig at that time and corresponds to the test duration used previously for similar uncoated alloys (6). In view of the similarity in general structural changes to the oxidation specimens described in detail above, discussion of the hot corrosion results will be limited to some observations having the most practical significance.

1. **Diffusion Aluminide Coatings**

These specimens remained intact at 300 h exposure except for certain alloys where corrosion penetration through the substrate had accelerated attack. The location of the penetration was generally at the interface with the cement bed. This indicated a possible reaction with the cement. However, the alloys effected had generally lower corrosion resistant compositions (i.e., in terms of Cr + Ta levels) that would be more sensitive to pitting and/or loss of the protective aluminide coating.

The higher substrate level of yttria tended to reduce, but not eliminate, porosity as observed in the oxidation studies. The comparison of alloys with different levels of aluminum was complicated by the presence of tantalum. Unlike alloy 51, most of the more complex alloys showed appreciable levels of porosity in the inner zone with the higher aluminum alloys generally tending to form less. Alloys with a more corrosion resistant base composition (i.e., 11 or 15% Cr) but a low aluminum level

such as INCONEL alloy MA 6000 could still be subject to coating breakdown
if the proportionately higher level of inner zone porosity linked with a
pit in the outer zone as shown in Figure 11. Substrate additions of
chromium + tantalum tended to increase the coating resistance to hot
corrosion penetration (Table V) but had no significant effect on porosity
formation.

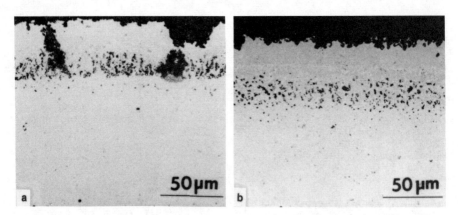

Figure 11. Comparison of aluminide coated (a) INCONEL alloy MA 6000 (4.5%
Al, 2.5% Ti) with (b) alloy 78 (7.0% Al, 0% Ti).

The addition of 5% Co (i.e., alloy 74) to alloy 51 tended, if at all,
to increase the propensity for porosity formation. Consequently, the hot
corrosion resistance of alloy 74 was slightly lower than alloy 51 because
of coating penetration (via outer zone pits linking with inner zone
porosity) as indicated in Table V.

2. Overlay Coatings

The EB-PVD and LPPS NiCoCrAlY coatings remained intact after 300 h
although some EB-PVD specimens showed end attack where the thickness of the
coating was insufficient.

Figure 12 shows that the lower yttria alloy 76 (0.6% Y_2O_3) formed more
porosity at the EB-PVD NiCoCrAlY coating-substrate interface than alloy 51
(1.1% Y_2O_3). However, the levels of porosity for both overlays were lower
than observed with aluminide coatings. In particular, the LPPS coating
showed the lowest levels with the layered structure possibly acting as a
vacancy sink. This effect is also observed with INCONEL alloy MA 6000
(Figure 13) with even less porosity associated with higher Al substrates
(alloy 78). The hot corrosion maximum attack results (Table V) indicate
(Cr+Ta) additions are beneficial particularly at the lower aluminum levels
and that the EB-PVD coating showed less penetration than the LPPS coating.
There was no beneficial effect of cobalt on the porosity formation or hot
corrosion resistance of the overlay coatings.

In general, the above observations on hot corrosion specimens
substantiated the oxidation test results. Substrate elements beneficial to
reducing porosity were effective in both aluminide and overlay coatings.
More elemental diffusion and hence porosity was found for the same
substrates with the aluminide coating which in hot corrosive environments

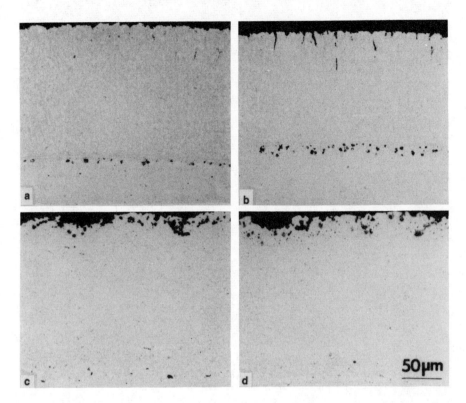

Figure 12. Comparison of EB-PVD NiCoCrAlY coated (a) alloy 51 (1.1% Y_2O_3), (b) alloy 76 (0.6% Y_2O_3) with LPPS NiCoCrAlY coated, (c) alloy 51, and (d) alloy 76 after 300h Burner Rig test.

could lead to a link-up of pits in the coating surface with porosity at the substrate-coating interface and consequent substrate attack. Further tailoring of coating and substrate composition together with the use of diffusion limitors suggests the potential for porosity-free coated ODS/MA components.

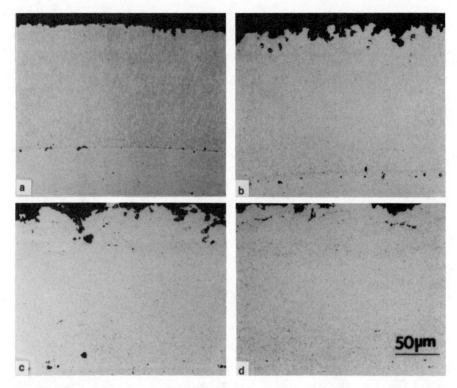

Figure 13. Comparison of EB-PVD NiCoCrAlY coated (a) INCONEL alloy MA 6000 (4.5% Al, 2.5% Ti), (b) alloy 78 (7% Al) with LPPS NiCoCrAlY coated, (c) MA 6000, and (d) alloy 78 after 300h Burner Rig test.

Conclusions

1. Diffusion aluminide coatings on all the ODS alloys tested exhibited porosity after 146 hours exposure at 1100°C while the overlay coatings did not show any porosity.

2. The presence of higher levels of aluminum in the substrate significantly reduced the formation of porosity in the aluminide coating.

3. Both overlay coatings on all the ODS alloys showed significantly better resistance to porosity formation when compared with the aluminide coating.

4. The presence of higher levels of aluminum in the substrate resulted in better resistance to porosity formation in the interdiffusion zone of the overlay coating produced by both processes.

5. The combined effect of chromium + tantalum in the substrate was beneficial for overlay and aluminide coatability even with lower levels of aluminum.

6. Decreasing levels of the dispersion strengthening phase Y_2O_3 had no beneficial effect in reducing the incidence of porosity formation.

7. There was no measurable beneficial effect of cobalt additions to the substrate alloy on porosity formation in aluminide or overlay coatings.

8. More elemental diffusion and hence porosity was found for the same substrates with the aluminide compared to the overlay coatings. In hot corrosive environments, this could lead to a link-up of pits in the coating surface with porosity at the substrate-coating interface and consequent substrate attack as observed on some aluminide coated alloys. The hot corrosion resistance of the overlay coated alloys was significantly better with no substrate attack.

9. Further tailoring of coating and substrate composition, together with the possible use of diffusion limitors, suggests the potential for porosity-free coated ODS components.

References

1. J. S. Benjamin, Metallurgical Transactions, Vol. 1, October, 1970, pp. 2943-2951.

2. Y. G. Kim and H. F. Merrick, NASA CR-159493, Contract No. NAS3-20093, NASA Lewis Research Center, May, 1979.

3. L. R. Curwick, Final Report, Contract No. N0019-75-C-0313, Naval Air Systems Command, December, 1976.

4. R. C. Benn, Final Report No. NADC-76204-30, Contract No. N-62269-76-C-0483, Naval Air Systems Command, December, 1977.

5. R. C. Benn, Final Report No. NADC-78096-69, Contract No. N62269-78-C-0200, Naval Air Systems Command, July, 1979.

6. R. C. Benn, Final Report No. NADC-79106-60, Contract No. N62269-79-C-0280, Naval Air Systems Command, May, 1981.

7. C. M. Austin and R. C. Benn, Final Report NADC-81109-60, Contract No. N62269-80-C-0298, Naval Air Systems Command, August, 1984.

8. C. M. Austin, Final Report No. NADC-80113-60, Contract No. N62269-81-C-0725, Naval Air Systems Command, August, 1984.

9. D. H. Boone, D. A. Crane, and D. P. Whittle, Thin Solid Films, Vol. 84, No. 1, pp. 39-48 (1981).

10. L. A. Monson and W. I. Pollock, Summary Report to Air Force Materials Laboratory, FTD, Contract AF33(615)-1704.

11. P. A. Bergman, C. T. Sims, and A. N. Beltran, ASTM STP 421, Am. Soc. Testing Materials, 1967, p. 43.

12. D. Manesse, Ch. Lopvet, and P. Mazars, Heurchrome, Marche Stpae No. 83.96.002.00.471.75.86, Rapport Final - Lot No. 2, August, 1985.

13. P. Huber, Sulzer, Private Communication, 1986.

14. R. Mervel, to be published, 1986.

15. P. Deb, D. H. Boone, and R. C. Benn, to be published.

ANALYSIS OF MICROSTRUCTURAL CHANGES DUE TO CYCLIC OXIDATION

IN ALUMINIDE-COATED Ni-Al, Ni-Cr, and Ni-Cr-Al ALLOYS

P. J. Fink* and R. W. Heckel

Department of Metallurgical Engineering
Michigan Technological University
Houghton, Michigan 49931

Abstract

Microstructural changes in the coating and adjacent substrate were studied in γ-phase, Ni-base alloys which had been given an aluminide diffusion coating and which were subsequently tested for cyclic oxidation resistance in air at 1100°C (one hour per cycle). Alloy substrate compositions were Ni-6.7Al, Ni-11.7Cr, Ni-22.6Cr, and Ni-11.5Cr-7.0Al (atomic percent). Oxidation behavior was characterized by type of oxide formed and weight change as a function of oxidation time. Changes in coating/substrate concentration-distance profiles, Cr and Al fluxes, and microstructures were compared to oxidation behavior. The results show that fluxes in the coating adjacent to the oxide scale correlated with spallation weight losses and oxide type. Coating/substrate fluxes were significant, and greatly dependent on substrate composition. Ternary diffusion "cross-term effects" were often large and, at times, dominant in determining the direction and magnitudes of both Cr and Al fluxes.

* Present address: General Electric Co., Evendale, Ohio

Introduction

The process of formation of diffusion aluminide coatings (pack cementation) has been analyzed extensively for Fe- and Ni-base substrates (1-9). These studies have shown the importance of interdiffusion in the gas phase surrounding the substrate and in the substrate itself during the formation of various aluminide layers. Comparatively little attention, however, has been given to the diffusional analysis of microstructural and compositional changes in aluminide-coated substrates which have undergone cyclic oxidation exposure.

The purpose of this research was to investigate the changes occurring in the microstructure and composition-distance profiles of Cr and Al in the coating and adjacent substrate region as a function of the duration of cyclic oxidation exposure. These changes were to be compared to oxidation parameters such as weight change and type of oxide formed as a function of time. Substrate compositions included a Ni-Al alloy, two Ni-Cr alloys, and a Ni-Cr-Al alloy (Table I); all compositions were within the Ni-rich, γ-phase, solid solution. The substrates were those studied in a companion research investigation of the cyclic oxidation degradation of low-pressure plasma-sprayed coatings (10-12), providing the opportunity for comparisons of the behavior of the two types of coating processes. In addition, selection of Ni-Cr-Al binary and ternary alloys allowed for determination of the Cr and Al fluxes in the γ phase because of the availability of ternary interdiffusion coefficients as a function of composition and temperature in the γ-phase (12).

Table I. Compositions of Alloy Substrates

Alloy Code	Nominal Composition (atomic percent) Cr	Al	Active Compositions (atomic percent) Cr	Al	(weight percent) Cr	Al
NiA	0	10	---	6.7	---	3.2
N1C	10	0	11.7	---	10.5	---
N2C	20	0	22.6	---	20.6	---
Ni1C1A	10	10	11.5	7.0	10.7	3.4

Experimental Details

The four substrates (Table I) were all given a standard diffusion aluminide processing treatment (Chromalloy; RT 21). Oxidation testing in air was carried out with one-hour cycles at 1100°C, each followed by a twenty-minute cooling to near room temperature. Weight change data as a function of oxidation exposure exhibited the usual initial increase in weight loss due to oxide spallation (Figure 1). The type of oxide remaining on the specimens was determined by x-ray diffractometry; the results are shown in Table II with a qualitative estimate of the relative abundance of each based on diffraction peak intensities.

Composition-distance profiles for Cr and Al were determined by energy dispersive analysis using a JEOL U35-C scanning electron microscope and a KEVEX-8000 X-ray microanalyzer. Compositionally-similar, Ni-Cr-Al γ-phase alloys of known compositions were used as standards; compositions by energy dispersive analysis were determined using an updated Magic V correction procedure, an LSI-11 minicomputer and a Z-8000 microprocessor.

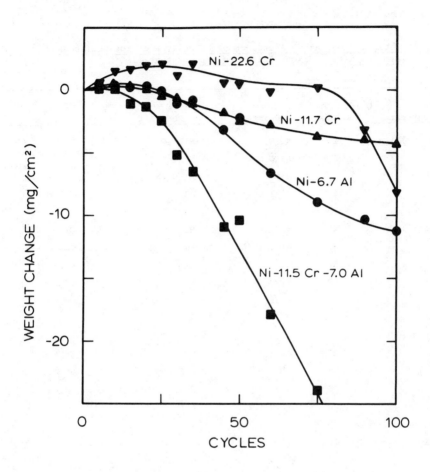

Figure 1 - Weight changes as a function of cyclic
oxidation exposure at 1100°C.

<u>Results</u>

Ni-6.7% Substrate

    The as-coated microstructure and the composition distance profile of
this substrate are shown in Figures 2 and 3. During oxidation Al diffused
to the outer surface of the specimen, causing the formation of the γ´
phase and, subsequently, the γ phase to the β layer is shown in Figure 4
after 35 cycles. Concentration-distance profiles for specimens oxidized
for up to 100 cycles are shown in Figure 5. As may be seen in Table II,
the transition to a non-protective NiO oxide occurred at 50 cycles and
significant spallation weight loss accompanied this transition (Figure 1).

Table II. Relative Abundance of Oxide Type on the Surfaces of Samples after Various Oxidation Exposures at 1100°C

Substrate	Oxidation Cycles	Primary	Moderate	Trace
Ni-6.7Al	25	Al_2O_3	$NiAl_2O_4$	NiO
	35	Al_2O_3	$NiAl_2O_4$	
	50	NiO	$Al_2O_3;NiAl_2O_4$	
	100	$NiAl_2O_4;NiO$	Al_2O_3	
Ni-11.7Cr	25	Al_2O_3	$NiAl_2O_4$	$Cr_2O_3;NiCr_2O_4$
	35	$Al_2O_3;NiCr_2O_4$		
	50	Al_2O_3	$NiAl_2O_4$	$NiCr_2O_4$
	75	Al_2O_3	$NiCr_2O_4$	$NiAl_2O_4$
	100	$NiCr_2O_4$	$Al_2O_3;NiAl_2O_4$	Cr_2O_3
Ni-22.6Cr	10	Al_2O_3	$NiAl_2O_4$	
	25	$Al_2O_3;NiAl_2O_4$	Cr_2O_3	$NiCr_2O_4$
	35	Al_2O_3	$NiAl_2O_4$	$Cr_2O_3;NiCr_2O_4$
	50	Al_2O_3	$NiAl_2O_4;NiCr_2O_4$	Cr_2O_3
	100	$Al_2O_3;Cr_2O_3$	$NiAl_2O_4;NiCr_2O_4;NiO$	
Ni-11.5Cr-7.0Al	10	Al_2O_3		
	25	Al_2O_3	$NiAl_2O_4;NiCr_2O_4$	$Cr_2O_3;NiO$
	35	$Al_2O_3;NiAl_2O_4$	$NiCr_2O_4;NiO$	
	50	Al_2O_3	$NiAl_2O_4;NiCr_2O_4$	
	75	$Al_2O_3;Cr_2O_3$	$NiAl_2O_4;NiCr_2O_4$	
	100	$Al_2O_3;Cr_2O_3;NiO$	$NiAl_2O_4;NiCr_2O_4$	

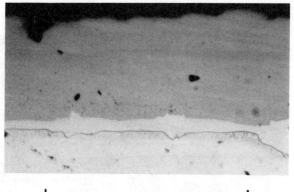

Figure 2 - Microstructure of the as-coated
Ni-6.7Al substrate; γ-phase substrate
(bottom), γ´ intermediate layer Ni$_3$Al), and
β phase (NiAl; grey) showing both low-Al and
high-Al regions.

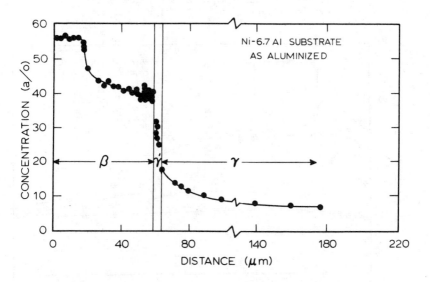

Figure 3 - Al concentration-distance profile for the
as-coated Ni-6.7Al substrate. Distance is measured
into the coating from coating/oxide interface. In
addition, interdiffusion between the coating and
substrate led to growth of the γ´ (Ni$_3$Al) layer,
solution of the β (NiAl) layer (and subsequently the
γ´ layer). The growth of the γ´ layer and solution
of the β layer is shown in Figure 4 after 35 cycles.
[Concentration-distance profiles for specimens oxidized
for up to 100 cycles are shown in Figure 5.]

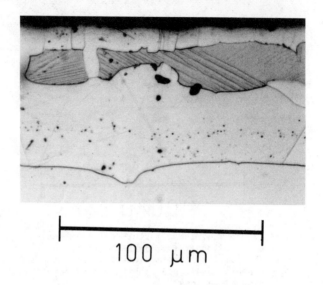

100 μm

Figure 4 - Coated Ni-6.7Al substrate of the
cyclic oxidation at 1100°C for 35 cycles.

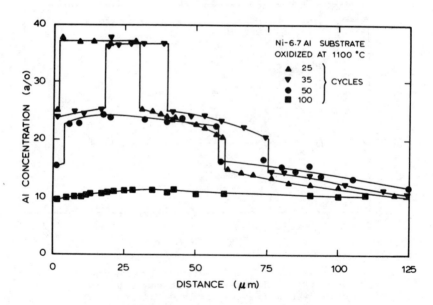

Figure 5 - Concentration-distance profiles for the
coated Ni-6.7Al substrate cyclically oxidized at
1100°C for 25, 35, 50, and 100 cycles. Distance is
measured from the surface of the coating (the oxide/
interface coating).

Ni-11.7Cr Substrate

The as-coated microstructure of this alloy is shown in Figure 6; the concentration-distance profiles for Cr and Al are shown in Figure 7. The $\beta + \alpha$ layer (α is the bcc phase which contains approximately 90 a/o Cr) is seen to have formed between the β (coating) and the (substrate) phases during aluminization due to the low solubility of Cr in β.

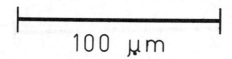

Figure 6 - As coated Ni-11.7Cr substrate.

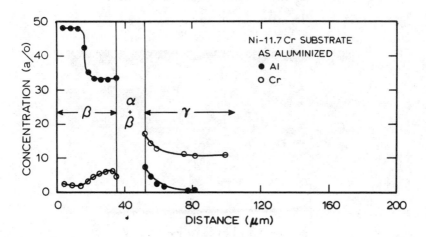

Figure 7 - Concentration-distance profiles for the as-coated Ni-11.7Cr substrate.

27

Oxidation of the coated Ni-11.7Cr substrate resulted in rapid solution of the α (with Cr enrichment of the underlying γ), and progressive solution of the β layer (and, subsequently, the γ′ layer) as shown in Figures 8 and 9. Significant Kirkendall porosity (relative to Figure 4) is evident in Figures 8 and 9. The concentration-distance profiles for Cr and Al are shown for the specimen oxidized for 25 cycles (Figure 8) in Figure 10. The enhancement of the Cr content in the γ phase beneath the coating due to the solution of the α phase is indicated by the maximum in the Cr concentration at a distance of about 120 μm from the oxide/coating interface.

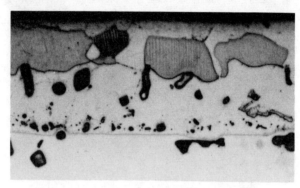

100 μm

Figure 8 - Coated Ni-11.7Cr after cyclic oxidation at 1100°C for 25 cycles; γ-phase substrate (bottom), γ′-phase layer, β-phase layer (discontinuous), and γ′ layer (at oxide/coating interface).

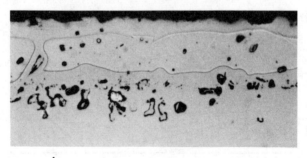

100 μm

Figure 9 - Coated Ni-11.7Cr after cyclic oxidation at 1100°C for 35 cycles; γ′-phase layer between γ-phase substrate, and γ-phase adjacent to oxide/metal interface.

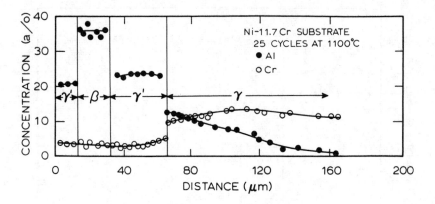

Figure 10 - Cr and Al concentration-distance
profiles for coated Ni-11.7Cr after 25
oxidation cycles at 1100°C.

Table II indicates that the formation of Cr-rich oxides (Cr$_2$O$_3$ and
NiCr$_2$O$_4$) developed relatively slowly during the oxidation of coated
Ni-11.7Cr substrate. The predominance of Al$_2$O$_3$ (and NiAl$_2$O$_4$) in the scale
is reflected in the low rate of weight loss due to spallation (Figure 1).

Ni-22.6Cr Substrate

Microstructures in both the as-coated and oxidized specimens having a
Ni-11.6Cr substrate were very similar to those of the Ni-11.7Cr substrate.
Kirkendall porosity formation was slightly more severe. Concentration/
distance profiles for the as-coated specimen and specimens oxidized for 25
and 50 cycles are shown in Figures 11, 12, and 13. The rapid solution of
the α phase and the resulting enrichment of the underlying γ phase in the
substrate is evident in Figure 12.

The transition to Cr-rich oxides for the Ni-22.6Cr substrate occurred
somewhat faster than for the Ni-11.7Cr substrate as shown in Table II. It
is somewhat anomalous that the rates of weight loss of these two coated
alloys are similar up to 75 cycles (Figure 1). However, the Ni-22.6Cr
alloy is seen to exhibit rapid weight loss due to oxide spallation after
75 cycles at which time significant amounts of Cr-rich oxides (and even
NiO) were found in the oxide scale.

Ni-11.5Cr-7.0Al Substrate

The microstructure of specimens oxidized (25 and 50 cycles) for the
Ni-11.5Cr-7.0Al substrate are shown in Figures 14 and 15. It is
significant that the α phase was very slow to dissolve during oxidation
exposure. Figure 14 shows that the α remained in the γ´ phase after
partial solution of the β phase. The low solubility of Cr in γ´ impeded
the diffusion of Cr to the γ-phase substrate. At longer times (Figure
15), the formation of γ in the layer which was previously γ´ + α is
obvious; the α and γ´ appear to have reacted to form the γ phase which has
a significantly higher solubility for Cr. The development of Kirkendall

29

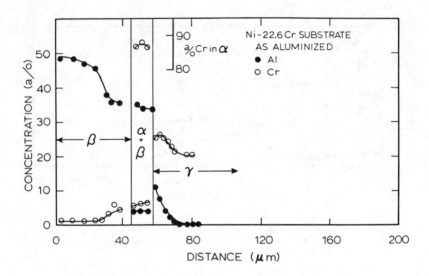

Figure 11 - Concentration-distance profiles in
the as-coated Ni-22.6Cr substrate.

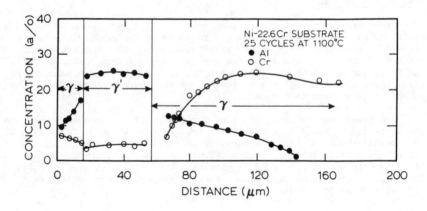

Figure 12 - Concentration-distance profiles in the
coated Ni-22.6Cr substrate after oxidation at 1100°C
for 25 cycles.

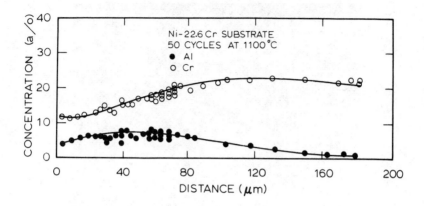

Figure 13 - Concentration-distance profiles in the
coated Ni-22.6Cr substrate of the oxidation at
1100°C for 50 cycles.

porosity was less severe for this substrate than for the Ni-Cr binary
alloy substrates.

The concentration-distance profiles for the as-coated and for the
oxidized specimens are shown in Figures 16, 17, 18, and 19. The retention
of the α phase in a γ´ matrix is shown to have retarded the movement of Cr
into the γ substrate; that is, no maximum appears in the Cr concentration
in the substrate after 25 cycles (Figure 17). Partial solution of the α
phase gave use to a slight maximum after 50 cycles (Figure 18). Continued
oxidation resulted in the loss of the α and γ´ phases (as well as the β)
as shown in Figure 19.

Figure 1 shows that the coated Ni-11.5Cr-7.0Al substrate exhibited an
extremely large rate of weight loss due to oxide spallation. The data in
Table II indicate that this alloy also developed Cr-rich oxides (and NiO)
very rapidly relative to the other substrates.

Discussion

Correlation Between Oxidation Data and Fluxes in Coatings

The oxidation data (Table II and Figure 1) shows reasonably good
agreement between rate of weight loss and type of oxide formed, as is
generally observed in cyclic oxidation studies; Al-rich oxide scales
generally have much better resistance to spallation than those containing
significant amounts of Cr-rich oxides or NiO. It was, therefore, of
interest to investigate a possible correlation between these oxidation
parameters and the fluxes of Cr and Al in the coating at the coating/oxide
interface.

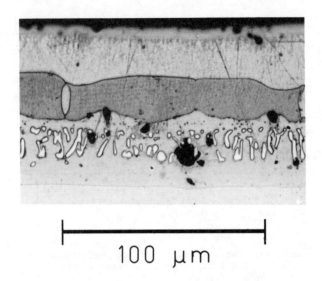

100 μm

Figure 14 - Coated Ni-11.5Cr-7.0Al after cyclic
oxidation at 1100°C for 25 cycles.

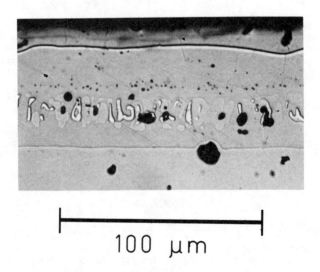

100 μm

Figure 15 - Coated Ni-11.5Cr-7.0Al after cyclic
oxidation at 1100°C for 50 cycles.

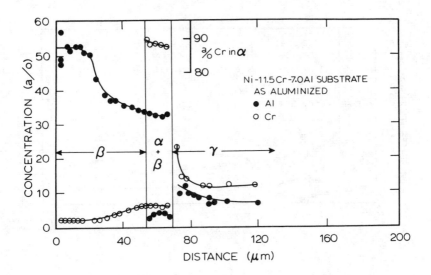

Figure 16 - Concentration-distance profiles for
as-coated Ni-11.5Cr-7.0Al.

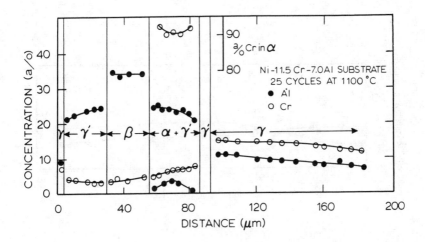

Figure 17 - Concentration-distance profiles for
coated Ni-11.5Cr-7.0Al after oxidation at 1100°C
for 25 cycles.

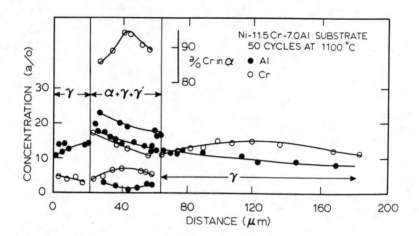

Figure 18 - Concentration-distance profiles for
coated Ni-11.5Cr-7.0Al after oxidation at 1100°C
for 50 cycles.

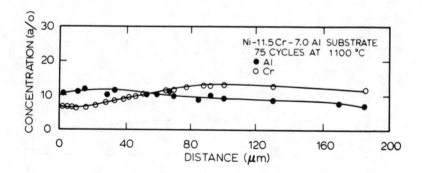

Figure 19 - Concentration-distance profiles for
coated Ni-11.5Cr-7.0Al after oxidation at 1100°C
for 75 cycles.

If the rate of weight loss during oxidation is due only to Al_2O_3 spallation, the slopes in Figure 1 (after dividing by the density of the alloy; approximately 9 gm/cm^3) yield the flux of Al from the oxidizing specimen. This will obviously be an "upper bound" for the Al flux, since the weight loss may include Cr and Ni, also. The fluxes of Cr and Al in the coating at the coating/oxide interface can be determined from

$$J_{Al} = - \tilde{D}_{AlAl} \cdot \frac{\partial C_{Al}}{\partial x} - \tilde{D}_{AlCr} \cdot \frac{\partial C_{Cr}}{\partial x} \tag{1}$$

$$J_{Cr} = - \tilde{D}_{CrAl} \cdot \frac{\partial C_{Al}}{\partial x} - \tilde{D}_{CrCr} \cdot \frac{\partial C_{Cr}}{\partial x} \tag{2}$$

where the J_i's are fluxes (a/o·cm/sec), the $\partial C_i/\partial x$'s are gradients (a/o/cm), and the $\tilde{D}$'s are the ternary interdiffusion coefficients (cm^2/sec). Equations 1 and 2 emphasize that the flux of each component is dependent upon two composition gradients and two interdiffusion coefficients. The availability of the ternary interdiffusion coefficients as a function of composition and temperature (in the range of 1100-1200°C) in the γ-phase of the Ni-Cr-Al system (12), therefore, allows the determination of the Cr and Al fluxes in the coating at the coating/oxide interface under conditions where the γ phase is adjacent to the oxide. Calculations made using Equations 1 and 2 assumed a composition of Ni-15Cr-5Cr for approximating the interdiffusion coefficients (cm^2/sec):

$$\tilde{D}_{AlAl} = 1.8 \times 10^{-10} \qquad \tilde{D}_{AlCr} = 0.3 \times 10^{-10}$$

$$\tilde{D}_{CrAl} = 0.8 \times 10^{-10} \qquad \tilde{D}_{CrCr} = 1.0 \times 10^{-10}$$

The fluxes obtained from weight change data and from Equations 1 and 2 are compared in Table III. Clearly, some scatter exists in the Table III fluxes calculated from Equations 1 and 2 due to the use of four experimentally determined parameters for each value. Comparison of weight change fluxes to coating fluxes indicates that, for the Ni-6.7Al substrate (100 cycles), the significant input of Ni into the oxide gives a higher weight change flux. Reasonable agreement is seen for the two types of calculations for the Ni-11.7Cr substrate which had the best oxidation (including spallation) resistance. The Ni-22.6Cr substrate appears to be somewhat anomalous in that the weight change fluxes (upper bound) are less than those calculated from gradients. However, the large weight gains in this alloy prior to the onset of significant oxide spallation would indicate that coating fluxes to the oxide would not correlate well with oxide spallation fluxes, especially for the short oxidation exposures. The Ni-11.5Cr-7.0Al fluxes indicate a significant difference in the two flux calculations, reflecting the large amount of Cr and Ni entering the oxide and being lost by spallation. In general, reasonable correlation existed between weight change data, type of oxide, and coating fluxes.

Coating/Substrate Interdiffusion

Flux analysis (Equations 1 and 2) of all four substrates indicates that both the Cr and Al fluxes in the substrate at the substrate/coating interface were very large in the as-coated specimens (i.e., the start of the first oxidation cycle). Fluxes in the range of 1×10^{-6} to 5×10^{-6} a/o·cm/sec for both Cr and Al were calculated. In general, these fluxes decreased markedly during the initial oxidations exposure.

35

Table III. Comparison of Fluxes Determined from the Rate of Weight Loss During Oxidation and from Cr and Al Gradients in the Coating at the Coating/Oxide Interface

Substrate Alloy	Oxidation Cycles	Flux (a/o·cm/sec)		
		Via Weight Loss*	Via Gradients	
			J_{Al}	J_{Cr}
Ni-6.7Al	100	-4×10^{-7}	-1×10^{-7}	----
Ni-11.7Cr	75	-2×10^{-7}	-3×10^{-7}	-1×10^{-7}
Ni-22.6Cr	25	-2×10^{-7}	-7×10^{-7}	-3×10^{-7}
	35	-2×10^{-7}	-2×10^{-7}	$+2 \times 10^{-8}$
	50	-2×10^{-7}	-4×10^{-7}	-2×10^{-7}
Ni-11.5Cr-7.0Al	35	-1×10^{-6}	-9×10^{-8}	$+1 \times 10^{-7}$
	50	-1×10^{-6}	-4×10^{-7}	-1×10^{-7}
	75	-1×10^{-6}	$+9 \times 10^{-9}$	-8×10^{-8}
	100	-1×10^{-6}	-6×10^{-8}	-5×10^{-8}

* Assuming all weight loss due to only Al_2O_3 spallation and, therefore, equivalent to the Al flux due to spallation.

To many interdiffusion "cross-term" effects (i.e., the effect of the Al gradient (and D_{CrAl}) on the flux of Cr, and vice versa) were very prominent for the substrates which contained Cr. For example, both Cr and Al had large inward fluxes in the Ni-11.7Cr substrate at 25 cycles (Figure 10) in spite of the maximum in the concentration-distance curve for Cr. (The negative Al gradient dominated the positive Cr gradient.)

For the Ni-22.6Cr substrate at 25 cycles (Figure 12), the very steep (positive) Cr gradient in the substrate adjacent to the coating gave rise to zero-flux planes for both Cr and Al in the vicinity of about 100 μm from the oxide coating interface. At shorter distances both Cr and Al fluxes are negative (toward the coating), the Cr gradient dominating the Al gradient in this region. At distances greater than 100 μm, both gradients become negative, producing inward (positive) fluxes of both Cr and Al. It is also of interest to note that, in the γ-phase layer adjacent to the oxide (Figure 12), the gradients have opposite signs. However, the Al gradient is dominant and the fluxes of both Al and Cr to the coating/oxide interface occur as indicated in Table III. This same effect is seen in the γ phase layer adjacent to the oxide on the Ni-11.5Cr-7.0Al substrate oxidized for 50 cycles (Figure 18); both the Al and Cr fluxes are directed toward the oxide.

Conclusions

The evolution of microstructures, composition-distance profiles, and diffusional fluxes have been characterized for the cyclic oxidation degradation of diffusion aluminide coated Ni-Al, Ni-Cr, and Ni-Cr-Al γ-phase alloys. These changes in the coating and adjacent substrate have been correlated with data obtained for weight change and for type of oxide formed as a function of time. Substrate composition played a major role in the oxidation behavior and the microstructural and compositional changes in the coating and adjacent substrate.

Fluxes of Cr and Al in the coating at the coating/oxide interface are consistent with the weight changes due to oxidation and spallation and with the type of oxide formed. Interdiffusion between the coating and the substrate was extensive and ternary interdiffusion "cross-term" effects were often significant.

Acknowledgements

The authors wish to thank the National Aeronautics and Space Administration (Lewis Research Center) for their support of this research through Grant NAG 3-244 and the Union Carbide Corporation for their support of this research through a fellowship grant. In addition, the authors greatly appreciate the assistance of NASA's Lewis Research Center for oxidation testing and the Chromalloy Corporation for applying the diffusion coatings. Numerous discussions with Michael A. Gedwill and J. A. Nesbitt during the progress of the research are gratefully acknowledged.

References

1. S. R. Levine and R. M. Caves, "Thermodynamics and Kinetics of Pack Aluminide Formation in IN-100", J. Electrochemical Soc., 121 (1974) 1051.

2. A. J. Hickl and R. W. Heckel, "Kinetics of Phase Layer Growth During Aluminide Coating of Nickel", Met. Trans. A, 6A (1975) 431.

3. B. K. Gupta, A. K. Sarkhel, and L. L. Seigle, "On the Kinetics of Pack Aluminization", Thin Solid Films, 39 (1976) 313.

4. R. W. Heckel, M. Yamada, C. Ouchi, and A. J. Hickl, "Aluminide Coating of Iron", Thin Solid Films, 45 (1977) 367.

5. H. C. Bhedwas, R. W. Heckel, and D. E. Laughlin, "Aluminide Coating on Directional γ´-δ Eutectics", Thin Solid Films, 45 (1977) 357.

6. B. K. Gupta and L. L. Seigle, "The Effect on the Kinetics of Pack Aluminization of Varying the Activator", Thin Solid Films, 73 (1980) 365.

7. N. Kandasaney, F. J. Pennis, and L. L. Seigle, "The Kinetics of Gas Transport in Halide Activated Aluminizing Packs", Thin Solid Films, 84 (1981) 17.

8. D. C. Tu and L. L. Seigle, "Kinetics of Formation of Aluminide Coatings on Ni-Cr Alloys", Thin Solid Films, 95 (1982) 47.

9. R. Sivakumar and L. L. Seigle, "On the Kinetics of the Pack Aluminization Process", <u>Met. Trans. A</u>, 13A (1982) 495.

10. J. A. Nesbitt and R. W. Heckel, "Modeling Degradation and Failure of Ni-Cr-Al Overlay Coatings", <u>Thin Solid Films</u>, 119 (1984) 281.

11. J. A. Nesbitt, B. H. Pilsner, L. A. Carol, and R. W. Heckel, "Cyclic Oxidation Behavior of β+γ Overlay Coatings on γ and γ + γ´ Alloys", <u>Superalloys 1984</u>, eds. M. Gell, C. S. Korbovich, R. Bricknell, W. B. Kent, and J. F. Radovich (Warrendale, PA: The Metallurgical Society, 1984) 701-710.

12. J. A. Nesbitt, "Overlay Coating Degradation by Simultaneous Oxidation and Coating/Substrate Interdiffusion" (NASA Technical Memorandum 83728, Lewis Research Center, 1984).

THE INFLUENCE OF VISCOUS FLOW ON THE KINETICS

OF GAS TRANSPORT IN ALUMINIZING PACKS

T.H.Wang and L.L.Seigle

Department of Materials Science and Engineering
College of Engineering and Applied Science
State University of New York at Stony Brook
Stony Brook, New York 11794

Abstract

   The combined effect of viscous flow and diffusion on the interface partial pressures and aluminum transport rate constants of halide-activated aluminizing packs has been calculated, using the "dusty gas model" for gas transport in a porous medium. The model permits an evaluation of the influence of pack particle size on the kinetics of aluminum transport, which, in chloride and fluoride activated packs, is shown to be appreciable only within the range of particle size 0.5 micron < r < 10 micron. This formulation is a generalization of the methods of calculation of Nciri and Vandenbulcke, and Levine and Caves , which represent limiting cases corresponding to gas transport in packs of very large and very small particle size respectively.

Introduction

Aluminum-rich coatings are extensively used to extend the life of superalloys in corrosive and oxidizing environments at high temperatures. The pack aluminizing process is commonly used to form aluminide coatings on such alloys (1-10). In this process aluminum is transferred from the bulk pack to the sample surface in the form of aluminum halide vapors, under the action of the thermodynamic activity gradient which exists between the pack and sample surface. The rate of coating formation is proportional to the rate of transport of these aluminum halide vapors through the pack.

Levine and Caves (1) calculated partial pressures of aluminum halide vapors and the aluminum transport rate constants in halide-activated aluminizing packs assuming : a) thermodynamic equilibrium between gas and condensed phases is attained at interfaces on both sides of an aluminum depleted zone between bulk pack and coating surface , b) transport of gaseous halides in the pack occurs by a purely diffusional mechanism. In applying Levine and Caves (L-C) method to an AlF_3-activated pack Kandasamy and Seigle (K-S) (7) showed that a difference in total pressure may exist between bulk pack and sample surface.

Recently Nciri and Vandenbulcke (N-V) (3) pointed out that the L-C and K-S methods neglected a bulk flow of the vapors, arising from the change in the total number of moles of gas due to reactions at the interfaces. They recalculated the aluminum transport rate constants for an NH_4Cl-activated pack using the same aluminum depleted zone model as K-S & L-C, but allowing for bulk flow as well as diffusion of the halide vapors in the pack. In their calculations they assumed that there is no difference in total pressure between pack and sample surface. Since flow of the halide vapors occurs in a porous medium , however , due to gas viscosity such a total pressure difference must exist. In this paper a more rigorous formulation of the aluminum transport rate in an aluminizing pack is given , which takes into account both the flow term suggested by N-V and the pressure drop between bulk pack and sample surface suggested by K-S. This formulation permits an evaluation of the effect of particle size on the kinetics of aluminum transport.

Model and Mathematical Formulation

The assumptions that our model uses for calculating partial pressures and the aluminum transport rate constant, Kg, are :
1) An aluminum depleted zone exists in the pack (1,2,10). An activator depleted zone (7), if such forms, is assumed to coincide with the aluminum depleted zone, as shown in Fig. 1.
2) The powder pack is modeled as a porous medium which consists of uniformly distributed spherical particles of equal diameter. This is the so-called "dusty gas model" (11,12,15,23).
3) Thermodynamic equilibrium is assumed to exist between gas and solid or liquid phases at interfaces on both sides of the depleted zone. The partial pressures of the gaseous species at these interfaces and the concentration of aluminum at the coating surface are assumed to be time-independent.
4) The main gaseous species in the pack is hydrogen. Other species are considered to interdiffuse with H_2 , i.e. , use of a protective atmosphere of H_2 is assumed.

The activators commonly used in the aluminizing pack can be divided into three classes 1) AlF_3, 2) Na(F, Cl, Br and I), all of which exist as condensed phases in the pack at normal operating temperatures , and 3)

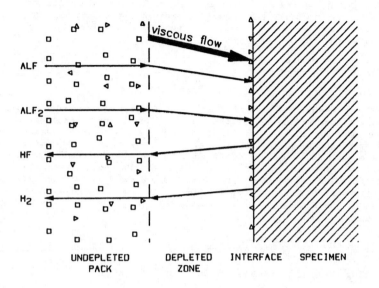

Figure 1 — Models of aluminum transport by the diffusional
flux and viscous flow in the depleted zone.

NH_4(Cl,Br and I), which do not form condensed phases at these temperatures. NH_4F or $NH_4F \cdot HF$ are equivalent to AlF_3 , since this will form in the
pack by reaction with aluminum powder when these activators are used.

The equilibrium partial pressures of aluminum halide vapors in the
bulk pack are calculated assuming that mass conservation is satisfied
within the pack and that the total pressure equals 1 atmosphere , as in the
papers of K-S & L-C (1,7) and Walsh (4). In calculating partial pressures
at the coating surface it is assumed that chemical reactions at the gas-
solid interface are rapid enough for a local chemical equilibrium to exist.

For the phenomenological conditions described above the total flux
for a given species is the sum of that due to bulk flow, which is necessary
to compensate for the change in the number of moles arising from chemical
reaction at the coating surface , and that due to diffusion, which occurs
because of the concentration gradient of gaseous species between two sides
of the depleted zone.   The total molar flux for a given species i , J_i ,
then can be written as (12)

$$J_i = J_i(\text{diffusion}) + J_i(\text{bulk flow})$$

For a real pack, the particle (or pore) size of the powder is large
enough for bulk flow to occur in the viscous regime, resulting in a pressure drop proportional to the flow rate.*   Assuming the pressure drop
----
* If the pore size of the powder pack is much smaller than the mean free
  path of the gas molecules then slip flux and Knudsen effects must be
  considered in the derivation (12).

41

across the depleted zone is small the density of the mixed gas will not vary appreciably with position in the depleted zone , and, as a first approximation , the gas can be treated as an incompressible fluid obeying Newton's law of viscosity (12,13,14). Therefore , the instantaneous total molar flux of species i , J_i , from bulk pack to the coating surface can be expressed as

$$J_i = \frac{\epsilon}{\ell RT} \left(D_i \frac{\Delta P_i}{d} + \frac{\bar{P}_i}{6 \pi r n_p f(\epsilon)\mu} \cdot \frac{\Delta P_T}{d} \right) \tag{1}$$

where D_i represents the interdiffusivity of species i in hydrogen , d is the depleted zone width, ΔP_i is the partial pressure difference of species i between the two sides of depleted zone, $\bar{P}_i$ is the average partial pressure of species i in the depleted zone, ΔP_T is the total pressure difference between the two sides of depleted zone , n_p is the number of powder particles per unit volume , r is the radius of the powder particle, μ is the coefficient of viscosity of the mixed gas, ϵ and ℓ are the porosity and tortuosity correction factors, and $f(\epsilon)$ is the influence function (15). The number of particles per unit volume , n_p , is

$$n_p = \frac{(1 - \epsilon)}{4/3 \ \pi r^3}$$

Substituting this expression for n_p into eq. (1), we get

$$J_i = \frac{\epsilon}{\ell RT} \left(D_i \frac{\Delta P_i}{d} + \frac{r^2 \bar{P}_i}{4.5(1 - \epsilon)f(\epsilon)\mu} \cdot \frac{\Delta P_T}{d} \right)$$

$$= \frac{\epsilon}{\ell RT} \left(D_i \frac{\Delta P_i}{d} + \frac{E \bar{P}_i}{\mu} \cdot \frac{\Delta P_T}{d} \right) \tag{2}$$

Where E is the "permeability" (12) defined by

$$E = \frac{r^2}{4.5(1 - \epsilon)f(\epsilon)}$$

For any given element, k, the flux of k through the depleted zone is

$$J_k = \sum_i C_{ki} J_i \tag{3}$$

where C_{ki} is the number of atoms of element k per molecule of species i.

Since the solubilities of hydrogen in the metals in which we are interested are very low (16,17), we can consider that no hydrogen enters the solid and the net flux of hydrogen at the coating surface , therefore , is zero :

$$J_H = \frac{\epsilon}{\ell RTd} [(2D_{H_2} \Delta P_{H_2} + D_{HX} \Delta P_{HX}) + \frac{E \ \Delta P_T}{\mu}(2\bar{P}_{H_2} + \bar{P}_{HX})] = 0 \tag{4}$$

where $\Delta P_{H_2} = \Delta P_T - \sum_j \Delta P_j$

42

and j represents all gaseous species except H_2. Substituting this expression for ΔP_{H_2} into eq. (4), we obtain

$$J_H = \frac{\epsilon}{\ell RTd} \left[2D_{H_2}(\Delta P_T - \sum_j \Delta P_j) + D_{HX} \Delta P_{HX} + \frac{E \Delta P_T}{\mu}(2\bar{P}_{H_2} + \bar{P}_{HX}) \right] = 0 \qquad (5)$$

Hence, after rearrangement

$$\Delta P_T [2D_{H_2} + \frac{E}{\mu}(2\bar{P}_{H_2} + \bar{P}_{HX})] + D_{HX} \Delta P_{HX} - 2D_{H_2} \sum_j \Delta P_j = 0 \qquad (6)$$

and

$$\Delta P_T = [2D_{H_2} \sum_j \Delta P_j - D_{HX} \Delta P_{HX}] / [2D_{H_2} + \frac{E}{\mu}(2\bar{P}_{H_2} + \bar{P}_{HX})] \qquad (7)$$

The instantaneous flux of aluminum through the depleted zone is

$$J_{Al} = \frac{\epsilon}{\ell RTd} [\sum_i C_{Al i} D_i \Delta P_i + A \sum_i C_{Al i} \bar{P}_i] \qquad (8)$$

and the instantaneous flux of aluminum which diffuses into the specimen is

$$F_{Al} = \frac{\epsilon}{\ell RTd} [\sum_i a_i C_{Al i} D_i \Delta P_i + A \sum_i a_i C_{Al i} \bar{P}_i] \qquad (9)$$

where a_i is the aluminum deposition coefficient of the ith species (5), and A is the flow velocity parameter, which can be written as

$$A = \frac{E}{\mu} \{ [2D_{H_2} \sum_j \Delta P_j - D_{HX} \Delta P_{HX}] / [2D_{H_2} + \frac{E}{\mu}(2\bar{P}_{H_2} + \bar{P}_{HX})] \} \qquad (10)$$

The aluminum transport rate constant, Kg, in the expression $W_{Al}^2 = Kg \cdot t$ is

$$Kg = 2 \rho M_{AL} F_{Al} d = \frac{2 \rho \epsilon M_{Al}}{\ell RT} [\sum_i a_i C_{Al i} D_i \Delta P_i + A \sum_i a_i C_{Al i} \bar{P}_i] \qquad (11)$$

where M_{Al} is the atomic weight of aluminum, ρ is the aluminum density in the pack, W_{Al} = grams of aluminum per cm^2, and t is the aluminizing time.

a) With AlF_3, NH_4F, and $NH_4F \cdot HF$ as activators, $AlF_3(s)$ will form in the pack . Therefore the partial pressure of $AlF_3(g)$ will equal the equilibrium vapor pressure of $AlF_3(s)$ and be independent of the aluminum activity. The independent reactions can be written for this case as :

$$AlF_3(s) = AlF_3(g)$$

$$2\underline{Al} + AlF_3(s) = 3AlF(g)$$

$$\underline{Al} + 2AlF_3(s) = 3AlF_2(g)$$

$$3/2H_2(g) + AlF_3(s) = \underline{Al} + 3HF(g)$$

$$2AlF_3(g) = Al_2F_6(g)$$

The partial pressures of the six significant gaseous species in the bulk pack can be determined from the above five equilibrium conditions , and the condition that the total pressure in the bulk pack is 1 atmosphere. The partial pressures of the six species at the coating surface are determined by the same five equilibrium conditions plus the kinetic condition of zero hydrogen flux at the surface.

b) For NaX activators the independent reactions are :

$$\underline{Al} + NaX = AlX(g) + Na(g)$$

$$\underline{Al} + 2NaX = AlX_2(g) + 2Na(g)$$

$$\underline{Al} + 3NaX = AlX_3(g) + 3Na(g)$$

$$NaX = NaX(g)$$

$$1/2H_2(g) + NaX = HX(g) + Na(g)$$

$$2AlX_3(g) = Al_2X_6(g)$$

where NaX signifies liquid or solid sodium-halide. The eight partial pressures in the bulk pack can be obtained from the six equilibrium conditions above , the Na - X balance ,

$$P_{Na}(g) = P_{AlX}(g)+2P_{AlX_2}(g)+3P_{AlX_3}(g)+6P_{Al_2X_6}(g)+P_{HX}(g)$$

and the assumption that

$$\sum_i P_i = 1 \text{ atm.}$$

The eight partial pressures at the coating surface can be determined by the six equilibrium conditions plus the kinetic conditions that at the coating surface

$$J_{Na} = J_X \quad \text{and} \quad J_H = 0$$

c) For NH_4X activators, where X is Cl, Br or I, it is assumed that dissociation of the activator occurs by a series of reactions , such as,

$$NH_4X(g) = NH_3(g) + HX(g)$$

$$NH_3(g) = 1/2N_2(g) + 3/2H_2(g)$$

that the final H : X ratio in the pack is 4 : 1, and that no condensed activator phase forms in the pack. The independent reactions are :

$$\underline{Al} + HX(g) = AlX(g) + 1/2H_2(g)$$

$$\underline{Al} + 2HX(g) = AlX_2(g) + H_2(g)$$

$$\underline{Al} + 3HX(g) = AlX_3(g) + 3/2H_2(g)$$

$$2AlX_3(g) = Al_2X_6(g)$$

$$\underline{Al} + 1/2N_2(g) = AlN(s)$$

The seven partial pressures in the bulk pack are determined by the above five equilibrium conditions, the condition that the H : X ratio is 4:1, leading to the equation

$$2P_{H_2}(g)+P_{HX}(g) = 4[P_{AlX}(g)+2P_{AlX_2}(g)+3P_{AlX_3}(g)+6P_{Al_2X_6}(g)+P_{HX}(g)]$$

and the condition that $\sum\limits_{i} P_i = 1$ atm.

The seven partial pressures at the coating surface are determined by the five equilibrium conditions and two kinetic conditions that at the coating surface

$$J_H = 0 \quad \text{and} \quad J_X = 0$$

After the partial pressures for all species in the bulk pack and at the coating surface are obtained, then from the general equation for the rate constant , eq. (11) , we are able to calculate the Kg values for individual cases.

Results and Discussion

Partial pressure and Kg calculations were carried out for packs with 4 wt.% pure aluminum powder (corresponding to an aluminum density, ρ = 0.04 g/cm^3). For packs activated with AlF$_3$, NH$_4$F·HF or NaX, in which a condensed activator phase exists, the amount of activator was in stoichiometric ratio with 4 wt% aluminum. For packs activated with NH$_4$(Cl, I), in which no condensed halide phase exists, the amount of activator was essentially that sufficient to produce a residual pack pressure of 1 atm. The balance of the pack was Al$_2$O$_3$ as inert filler. Values of equilibrium constants for the various reactions were obtained with the aid of thermodynamic data from the JANAF tables (22). Other parameters used for the calculations were porosity , ϵ = 0.7, tortuosity, ℓ = 4,(5,7) and influence function, f(0.7)=1/0.09819 , taken from Neale and Nader's paper (15).

The interdiffusional coefficients, D_i, of the gaseous species with H$_2$ were estimated using a Gilliland-type semi-empirical equation (18):

$$D_i = \frac{0.043 \cdot [T^3 \cdot (1/M_i + 1/M_{H_2})]^{1/2}}{P_T[V_i^{1/3} + V_{H_2}^{1/3}]}$$

where M_i and V_i are molecular weight and diffusional molar volume at the normal boiling point of species i , and M_{H_2} and V_{H_2} are the molecular weight and diffusional molar volume of H$_2$ (19,20). P_T is total pressure in the system. The Le Bas' additivity rule was applied to calculate the diffusional molar volume :

$$V_i = \sum\limits_{p} V_p$$

where V_p is the molar volume of element p in species i at the normal boiling point (19,20). V_{Al} is extrapolated from the density of liquid aluminum at the normal boiling point (21).

An iterative calculating process was used to obtain self-consistent values for partial pressures, total pressures and Kg. Calculations were done for AlF$_3$, NaF, NaCl, NH$_4$Cl and NaI activators from 800°C to 1093°C and surface aluminum activity between 10^{-4} and 0.5. The results are shown as

45

three dimensional plots in Figs. 2, 3, 4, 5 and 6. As seen in Fig. 2 the variation of Kg with particle radius, r, is S-shaped with the value of Kg approaching a lower limit as r approaches zero, and an upper limit as r approaches infinity. The lower limit is that value obtained when the bulk flow term is neglected, but a total pressure difference between bulk pack and sample surface is assumed to exist , as in the calculations of K-S (7). The upper limit is that value obtained when the bulk flow term is taken into account , but the total pressure difference between bulk pack and sample surface is neglected , as in the calculations of N-V (3).

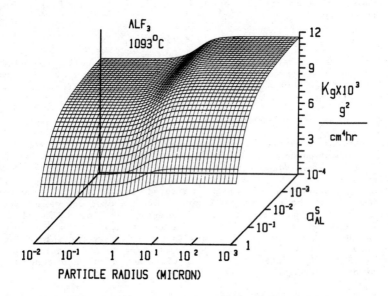

Figure 2 - Plot of Kg vs. particle radius and coating surface aluminum activity for the AlF_3-activated pack at $1093^{\circ}C$.

Figs. 3, 4, 5 and 6 are plots of Δ Kg vs. particle radius , coating surface aluminum activity and temperature for AlF_3 and NH_4Cl-activated packs. ΔKg is the relative change of rate constant defined as

$$\Delta Kg(\%) = \frac{Kg(present\ calculation) - Kg(\ K\text{-}S\ method\ )}{Kg(\ K\text{-}S\ method\ )} \times 100$$

Values of particle radius, r, ranged from 0.01 micron to 1000 microns, coating surface aluminum activity from 0.5 to 10^{-4}, and temperature from $800^{\circ}C$ to $1093^{\circ}C$. The influence of r on ΔKg varies with temperature for a specified coating surface aluminum activity, or with coating surface aluminum activity for a specified temperature. As seen from Figs. 3 and 4 the effect of r on ΔKg is smaller at low temperatures than at high temperatures.   For aluminum and sodium halide activators, the effect of r on ΔKg at a constant temperature is always smaller at low coating surface aluminum activities than at high activities, as shown in Fig. 5. For ammonium halide activators, a reverse characteristic was found, as seen from Fig. 6.   The maximum values of ΔKg in the temperature range $800^{\circ}C$ to $1093^{\circ}C$ varied from 0% to 5% for chloride activators and from 0% to 20% for fluoride activators.

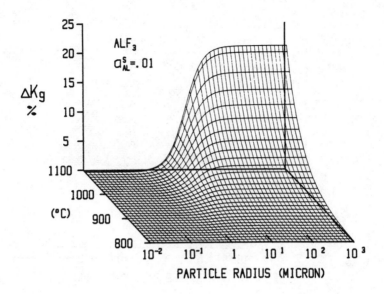

Figure 3 - Plot of ΔKg vs. particle radius and temperature for the AlF$_3$-activated pack at coating surface aluminum activity 0.01.

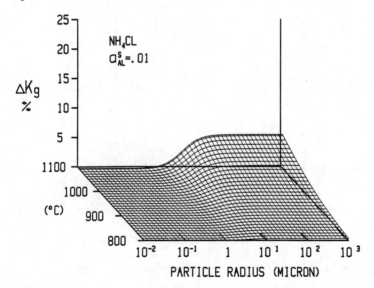

Figure 4 - Plot of ΔKg vs. particle radius and temperature for the NH$_4$Cl-activated pack at coating surface aluminum activity 0.01.

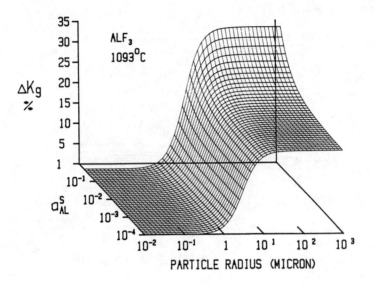

Figure 5 - Plot of ΔKg vs. particle radius and coating surface aluminum activity for the AlF$_3$-activated pack at 1093°C.

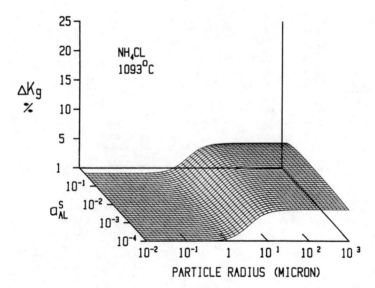

Figure 6 - Plot of ΔKg vs. particle radius and coating surface aluminum activity for the NH$_4$Cl-activated pack at 1093°C.

Some calculated results for all the activators at 1093°C and coating surface aluminum activity 0.01 are listed in Tables I, II, III and IV. In Tables I and II the effect of varying particle radius, r, on total pressures and Kg values is shown. It is seen that the values of Kg from the present method, which takes into account both viscous flow and total pressure difference, are always between the Kg values obtained by the K-S, L-C method (r = 0 micron), and the N-V method (r = ∞ micron). The same is true for surface total pressures.

Table I. Kg of different activators at 1093°C and surface aluminum activity 0.01

Particle size (micron)	Kg (g^2/cm^4hr)				
	AlF_3 ($\times 10^{-2}$)	NH_4Cl ($\times 10^{-3}$)	NaF ($\times 10^{-3}$)	NaCl ($\times 10^{-4}$)	NaI ($\times 10^{-4}$)
K-S method	0.9158	5.0790	5.3115	2.4259	2.2316
0.01	0.9158	5.0790	5.3115	2.4259	2.2316
0.1	0.9160	5.0793	5.3136	2.4259	2.2316
0.5	0.9207	5.0850	5.3620	2.4273	2.2319
1	0.9340	5.1016	5.4871	2.4308	2.2327
5	1.0596	5.2763	6.1521	2.4459	2.2457
10	1.0958	5.3398	6.2589	2.4480	2.2532
50	1.1112	5.3702	6.2989	2.4488	2.2577
1000	1.1118	5.3708	6.3006	2.4488	2.2580
N-V method	1.1118	5.3957	6.3006	2.4488	2.2580

Table II. Surface total pressures of all species at 1093°C and surface aluminum activity 0.01

Particle size (micron)	Total pressure at coating surface (atm.)				
	AlF_3	NH_4Cl	NaF	NaCl	NaI
K-S method	0.7905	0.9214	0.8405	0.9920	0.9920
0.01	0.7905	0.9292	0.8405	0.9920	0.9920
0.1	0.7906	0.9293	0.8408	0.9920	0.9920
0.5	0.7951	0.9306	0.8479	0.9925	0.9921
1	0.8078	0.9345	0.8667	0.9937	0.9923
5	0.9389	0.9762	0.9741	0.9990	0.9963
10	0.9807	0.9920	0.9926	0.9997	0.9986
50	0.9992	0.9996	0.9997	1.0000	0.9999
1000	1.0000	1.0000	1.0000	1.0000	1.0000
N-V method	1.0000	1.0000	1.0000	1.0000	1.0000

As seen from Tables III and IV, the variations of the partial pressures with r are similar to those of Kg for NaF and NH_4Cl activators, and are similar in shape but reversed in tendency for NaCl and NaI activators. Partial pressures of aluminum fluorides in the pack activated with AlF_3 do not vary with r. The explanation of these differences is connected with the details of the thermodynamic equilibria established in the presence of the various activators.

Table III. Surface partial pressures of AlX species at $1093^{\circ}C$ and surface aluminum activity 0.01

Particle size (micron)	P_{AlX} at coating surface (atm.)				
	AlF_3 ($\times 10^{-3}$)	NH_4Cl ($\times 10^{-3}$)	NaF ($\times 10^{-3}$)	NaCl ($\times 10^{-4}$)	NaI ($\times 10^{-4}$)
K-S method	9.5285	7.9900	4.7218	2.3036	1.8683
0.01	9.5285	7.9900	4.7218	2.3036	1.8683
0.1	9.5285	7.9903	4.7219	2.3035	1.8683
0.5	9.5285	7.9979	4.7258	2.3031	1.8682
1	9.5285	8.0195	4.7360	2.3020	1.8679
5	9.5285	8.2526	4.7990	2.2971	1.8625
10	9.5285	8.3393	4.8107	2.2964	1.8593
50	9.5285	8.3811	4.8152	2.2961	1.8575
1000	9.5285	8.3825	4.8155	2.2960	1.8574
N-V method	9.5285	8.4175	4.8155	2.2960	1.8574

Table IV. Surface partial pressures of AlX_2 species at $1093^{\circ}C$ and surface aluminum activity 0.01

Particle size (micron)	P_{AlX_2} at coating surface (atm.)				
	AlF_3 ($\times 10^{-3}$)	NH_4Cl ($\times 10^{-2}$)	NaF ($\times 10^{-4}$)	NaCl ($\times 10^{-6}$)	NaI*
K-S method	3.8261	0.9045	9.3945	7.5182	-
0.01	3.8261	0.9045	9.3945	7.5182	-
0.1	3.8261	0.9046	9.3951	7.5181	-
0.5	3.8261	0.9063	9.4104	7.5151	-
1	3.8261	0.9112	9.4513	7.5079	-
5	3.8261	0.9649	9.7043	7.4759	-
10	3.8261	0.9853	9.7516	7.4714	-
50	3.8261	0.9952	9.7699	7.4697	-
1000	3.8261	0.9955	9.7712	7.4696	-
N-V method	3.8261	1.0039	9.7712	7.4696	-

* Data are not available for this activator.

To complete the discussion, the results of the calculations described in this paper can briefly be explained as follows: The treatment of L-C and K-S, which neglects the flow term but allows for a difference in total pressure between bulk pack and coating surface , is valid for a pack with infinite resistance to viscous flow , such as a pack with infinitesimally small pore (powder-particle) size. Neglect of the flow term makes K_g of the L-C , K-S method come out on the low side. The treatment of N-V , which allows for the flow term but neglects the total pressure difference between bulk pack and coating surface , corresponds to a pack with zero resistance to viscous flow , such as a pack with an infinitely large pore (powder-particle) size. Neglect of the pressure drop due to friction accompanying viscous flow makes K_g of the N-V method come out on the high side. Aluminum transport rate constants from the two methods are, however, not greatly different.

Varying particle radius, r, corresponds to varying the pore size while keeping pack porosity (or packing density) constant in this calculation. Therefore , the results illustrate how the particle size of the pack powder influences the equilibrium partial pressures of gaseous species in the pack, and the aluminum transfer rate constant, K_g. The effect of particle size is appreciable only when r is between .5 and 10 micron. For $r < .5$ micron the resistance to viscous flow is so great that the flow velocity becomes negligible. For $r > 10$ micron the resistance to viscous flow is so small that the drop in total pressure between bulk pack and specimen surface becomes negligible.

Summary and Conclusions

A more rigorous calculation of equilibrium partial pressures and the kinetics of gaseous transport in halide activated aluminizing packs has been formulated , using the "dusty gas model" to simulate the powder pack. This treatment considers mass transfer effects from both diffusional and viscous flows , and allows for the existence of a difference in total pressure between bulk pack and coating surface . Values of partial pressures ,aluminum transport rate constants , K_g , and relative changes of K_g in pure aluminum packs have been calculated for various pack particle sizes , coating surface aluminum activities , and temperatures , while keeping packing density constant. We can conclude that :

1) The K_g and partial pressure values obtained by this method fall between those of the N-V, and the K-S & L-C methods , which represent limiting values corresponding to powders of infinitely large and infinitely small particle size , respectively.

2) The effect of powder particle size on K_g and partial pressures is greater at high temperatures than at low. For aluminum halide and sodium halide activators , the effect is greater at high aluminum activities at the coating surface , while for ammonium halide activators it is greater at low coating surface aluminum activities.

3) The influence of particle size on K_g is appreciable only for $.5$ micron $<$ $r < 10$ micron. When $r < .5$ micron the flow velocity is negligible while for $r > 10$ micron the drop in total pressure between bulk pack and sample surface is negligible.

4) The values of aluminum transport rates calculated by the N-V and L-C, K-S methods differ by only a few percent in most cases. At constant packing density , powder-particle size has a relatively small influence on the kinetics of aluminum transport in the pack.

References

1. S. R. Levine and R. M. Caves, "Thermodynamics and Kinetics of Pack Aluminide Coating Formation on IN-100," J. Electrochem. Soc., 121 (1974), 1051-1064.

2. H. Brill-Edwards and M. Epner, "Effect of Material Transfer Mechanisms on The Formation of Discontinuities in Pack Cementation Coatings on Superalloys," Electrochem. Tech., 6 (1968), 299-307.

3. B. Nciri and L. Vandenbulcke, "Influence Relative des Transports en Phase Gazeuse et dans Le Solide sur L'aluminisation du Fer et des Aciers," J. Less Common Metals, 95 (1983), 191-203.

4. P. N. Walsh, "Chemical Aspects of Pack Cementation," Chemical Vapor Deposition, 4th. International Conference, ed. G. F. Wakefield and J. M. Blocher (Princeton, NJ: The Electrochemical Society, 1973), 147-168.

5. B. K. Gupta, A. K. Sarkhel and L. L. Seigle, "On The Kinetics of Pack Aluminization," Thin Solid Films, 39 (1976), 313-320.

6. N. B. Arzamasov and D. A. Prokoshin, "Theoretical Problems of Diffusion Metallization from Halide Gaseous Media," Protective coatings on Metals, Vol. 5, ed. G. V. Samsonov, Translated from Russian by Erric Renner, (New York, NY: CONSULTANT BUREAU, 1973), 43-47.

7. N. Kandasamy, L. L. Seigle and F. J. Pennisi, "The Kinetics of Gas Transport in Halide-activated Aluminizing Packs," Thin Solid Films, 84 (1981), 17-27.

8. G. H. Marijnissen, "Codeposition of Chromium and Aluminum during a Pack Process," High Temperature Protective Coatings, ed. S. C. Singhal, (Warrendale, PA: The Metallurgical Society of AIME, 1983), 27-35.

9. R. Sivakumar and L. L. Seigle, "On The Kinetics of The Pack-aluminization Process," Metall. Trans. A, 7A (1976), 1073-1079.

10. B. K. Gupta and L. L. Seigle, "The Effect on The Kinetics of Pack Aluminization of Varying The Activators," Thin Solid Films, 73 (1980), 365-371.

11. R. B. Evans III, G. M. Watson and E. A. Mason, "Gaseous Diffusion in Porous Media at Uniform Pressure", J. Chem. Phys., 35 (1961) 2076-2083.

12. R. E. Cunningham and R. J. J. Williams, Diffusion in Gases and Porous Media, 1st ed. (New York, NY: Plenum Press, 1980), 129-168.

13. Richard H. F. Pao, Fluid Dynamics, 1st ed. (Columbus, Ohio: Charles E. Merrill Books Inc., 1967), 267-306.

14. J. C. Slattery, "Single-phase Flow Through Porous Media," A.I.Ch.E. Journal, 15 (1969), 866-872.

15. G. H. Neale and W. K. Nader, "Prediction of Transport Processes within Porous Media: Creeping Flow Relative to a Fixed Swarm of Spherical Particles," A.I.Ch.E. Journal, 20 (1974), 530-538.

16. Sung-man Lee and Jai-young Lee, "The Trapping and Transport Phenomena of Hydrogen in Nickel," Metall. Trans. A, 17A (1986), 181-187.

17. E. W. Johnson and M. L. Hill, "The Diffusivity of Hydrogen in Alpha Iron," Transaction TMS-AIME, 218 (1960), 1104-1112.

18. E. R. Gilliland, "Diffusion Coefficients in Gaseous Systems," Ind. Eng. Chem., 26 (1934), 681-685.

19. T. Hobler, Mass Transfer and Absorbers, 1st ed. (Long Island City, NY: Pergamon Press, 1966), 74-79.

20. W. Jost, Diffusion in Solids, Liquids and Gases, 3rd Printing, (New York, NY: Academic Press Inc., 1960), 406-435.

21. E. A. Brandes, ed., Smithell Metal's Reference Book, 6th ed. (Bodmin, Cornwall: Robert Hartnoll Ltd., 1983), (14-6)-(14-10).

22. D. R. Stull et al., eds. JANAF Thermochemical Tables, 2nd ed. (Midland, Michigan: Dow Chemical Co., U.S. Government Printing Office, 1971); and JANAF Thermochemical Tables Supplements (1974 and 1978).

23. E. A. Mason, A. P. Malinauskas and R. B. Evans III, "Flow and Diffusion of Gases in Porous Media," J. Chem. Phys., 46 (1967), 3199-3216.

KINETICS OF COATING/SUBSTRATE INTERDIFFUSION IN MULTICOMPONENT SYSTEMS

M. S. Thompson and J. E. Morral

Department of Metallurgy, Institute of Materials Science
U-136, University of Connecticut, Storrs, CT   06268

Abstract

    Three stages are identified for the kinetics of a multicomponent
alloy coating diffusing into its substrate. The first stage resembles
diffusion couple behavior and it continues until the reaction zone ex-
pands to the surface of the coating. The second stage is the most complicated of the three stages and its kinetics cannot be predicted without
referring to a particular alloy system.   The third and final stage resembles the late stages of homogenization and it refers to the asymptotic
approach of the coating to its final composition. The three stages are
illustrated with the example of interdiffusion between a single phase
coating and a substrate with which it is isomorphous. In this case the
kinetic constants are simple functions of the square root diffusivity.
As an illustration the results are applied to gamma phase, Ni-Cr-Al
systems.

Introduction

Interdiffusion can severely limit the service life of high temperature coatings. The phenomenon can result in loss of oxidation resistance of the coating as well as a reduction in the mechanical properties of the substrate alloy. An example is the high temperature degradation of MCrAlY coatings applied to turbine engine hardware. Several approaches used to control the adverse effects of interdiffusion include making the coatings thicker (1,2), applying diffusion barriers (3) and adjusting the coating chemistry in order to minimize the kinetics of interdiffusion. The latter approach is the topic of the following work. Of specific interest is developing analytical models that can be used to design coatings that resist interdiffusion.

Most previous theoretical descriptions of interdiffusion have applied only to binary systems. In these systems, interdiffusion is governed by a single parameter for each phase, the diffusivity, given by the symbol "D". In multicomponent systems, interdiffusion is governed by an array of parameters given by the diffusivity matrix, $[D]$. For example, in a ternary phase the diffusivity matrix consists of four D_{ij} coefficients. One problem with equations that describe multicomponent diffusion kinetics is that they tend to be long and complicated expressions which are functions of the $[D]$ coefficients.

Recently it has been shown that many multicomponent diffusion equations can be simplified by making use of the square root diffusivity matrix, $[r]$ (4-7). This matrix is related to $[D]$ by the expression

$$[D] = [r]\ [r] \qquad\qquad (1)$$

The $[r]$ matrix is the same size as the diffusivity matrix and each coefficient r_{ij} has a simple interpretation. It relates the interdiffusion of component i to the composition difference of component j across a diffusion couple (i.e. it is not related to concentration gradients as are the $[D]$ coefficients). However, it is important to appreciate that each r_{ij} is not the square root of the corresponding coefficient D_{ij}. The relationship is far more complex (6) and it is for this reason that substituting $[D]$ for $[r]$ has such a complicating effect.

With simplified versions of the multicomponent diffusion equations, it is possible to answer a number of questions relevant to the design of interdiffusion resistant coatings. For example questions like

56

1. How does the rate of interdiffusion vary versus the concentration of components in the coating?

2. To what extent can small changes in coating chemistry change coating life?

3. How does the initial rate of interdiffusion compare with the interdiffusion rate during the later stages of coating life?

can be answered quantitatively if there is both an appropriate kinetic model and experimental diffusivity data available. In the present report a model will be given for the case of a single phase coating which diffuses into an isomorphous substrate and it will be applied to the Ni-Cr-Al system. However before discussing the model, several general ideas of how to measure the extent of interdiffusion and analyze data will be considered.

Measures of Interdiffusion

The extent to which interdiffusion has occurred can be measured in a number of ways, for example by measuring the size of the reaction zone. For the case of high temperature coatings two relevant parameters are the chemical composition at the coating surface, for it determines oxidation resistance, and the amount of solute that has left or entered the coating, for it determines the extent to which substrate properties are affected. Figure 1 is a schematic representation of these parameters for a ternary system. The concentration of each solute at the surface is given by C_i^s

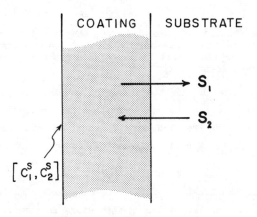

Figure 1 - Parameters for interdiffusion studies in ternary systems: the solute concentrations at the surface, C_1^s and C_2^s, and the amounts solute leaving or entering the coating, S_1 and S_2.

while the amount of each solute lost (or gained) in time t is given by S_i. In this work the fraction of solute leaving (or entering) the coating will be considered in order to separate the effects of coating chemisty from the effects of coating thickness. It is given by:

$$f_i = S_i/S_i^\infty \qquad (2)$$

in which S_i^∞ is the total solute which leaves the coating in infinite time.

Three stages of Interdiffusion

Three stages of interdiffusion will normally be seen in plots of either surface concentration versus time or fraction of solute lost versus time, as shown in Figure 2. In stage I, the reaction zone is less than the coating thickness. During this time, the coating/substrate

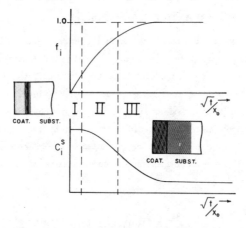

Figure 2 - The fraction of solute i, f_i, that has left or entered the coating and the surface concentration of i, C_i^s, plotted versus the square root of time. The time variable has been adjusted to account for the initial coating thickness, x_o.

system behaves like an infinite diffusion couple. Characteristics of this stage are that the surface concentration remains constant at the initial value and the fraction of solute lost changes linearly with the square root of time. The latter relationship can be stated as

$$f_i = k_i^I \ (\sqrt{t}/x_o) \qquad (3)$$

in which k_i^I is the kinetic constant for component i in stage I. If experimental f_i data collected during this stage is plotted versus $\sqrt{t}/x_o$, it should be possible to compare samples that had a different initial coating thickness (1).

In stage II the behavior depends on the specific system involved and the kinetics cannot be generalized as it could for stage I. The nucleation of new phases or the loss of initial phases can make this stage particularly difficult to model.

In stage III the system behaves like a thin homogeneous film which is dissolving into its underlying substrate. The kinetic behavior expected is that f_i will approach a value of one asymptotically (i.e. the solute in the coating will approach the equilibrium value in infinite time), while the surface concentration decays to its final value. An equation for the final stage kinetics which is followed by simple models is

$$f_i = 1 - k_i^{III} \, (\sqrt{t}/x_o)^{-1} \qquad (4)$$

in which k_i^{III} is the kinetic constant for interdiffusion of solute i during stage III.

Equations (3) and (4) provide a basis for evaluating the effect of alloy modifications on the interdiffusion of coatings. Improvements during stage I with respect to component i are indicated by a decrease in the value of k_i^I, while improvements during stage III are indicated by an increase in the value of k_i^{III}. Therefore, comparisons between different coating/substrate systems can be made based on the numerical values of the kinetic constants.

Effect of Composition on Interdiffusion

Analytical expressions for k_i^I and k_i^{III} can be derived which depend on the number and the distribution of phases present in the reaction zone and these can provide a basis for the design of interdiffusion resistant coatings. As an illustration the case of coating and substrate that are isomorphous and have constant diffusivity will be considered for the remainder of the present work. Figure 3 gives the boundary conditions of the isomorphous case. At time t=0, the concentration profile is a step function where the initial concentration of each solute in the coating is given as C_i^o and the concentration in the substrate as C_i^b. After time $t=t_1$, a new profile can be calculated. The amount of solute which has left the coating is equal to the area of the shaded region, S_i. This area can be determined by integrating the concentration profile as shown. Then the fraction of solute lost can be calculated from

$$f_i = S_i/\Delta C_i x_o \qquad (5)$$

in which ΔC_i is the difference in concentration between the coating and

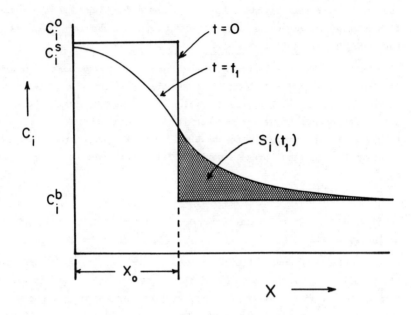

Figure 3 - The initial and subsequent composition profile for the example of an isomorphous coating and substrate. S_i is the amount of solute which has left the coating.

the substrate $(c_i^o - c_i^b)$ and $\Delta C_i x_o$ is the total solute which will leave (or enter) the coating in infinite time. In cases where there is no difference in concentration for one or more components (an unlikely possibility) equation (5) contains a singularity. In this case there will be a transient exchange of solute between the coating and substrate, which can be expressed in terms of the variation of S_i with time.

Crank (8) has discussed the case of binary interdiffusion in detail. His error function solution yields the plot given in Figure 4 of fraction solute lost versus time. On the figure the general equations proposed for stages I and III are compared with the analytical solution from Crank and it can be seen that there is good agreement. Indeed, Crank's solution reduces to these equations in the limits of early and late times of interdiffusion.

The two kinetic constants taken from the limiting equations are

$$k^{I} = \sqrt{D}/\sqrt{\pi} \tag{6}$$

and

$$k^{III} = (\sqrt{D})^{-1}/\sqrt{\pi} \tag{7}$$

60

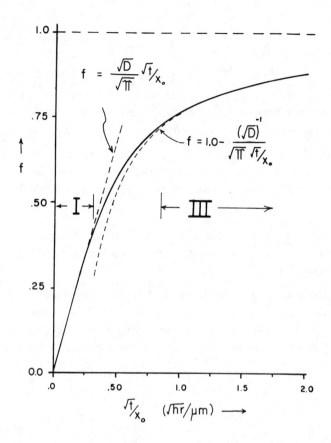

$$f = \frac{\sqrt{D}}{\sqrt{\pi}} \frac{\sqrt{t}}{x_o}$$

$$f = 1.0 - \frac{(\sqrt{D})^{-1}}{\sqrt{\pi}} \frac{\sqrt{t}}{x_o}$$

Figure 4 - Fraction of solute leaving a coating versus the square root of time for a binary, isomorphous system (solid line) compared to the limiting behavior predicted for stages I and III of interdiffusion (dashed lines).

Equations (6) and (7) show no explicit dependence of the constants on composition. Any composition dependence results implicitly from the effect of composition on D. Also, the equations show that the early and late stage kinetics are related. If interdiffusion is reduced in the stage I (i.e. if k^I is reduced) then it follows that interdiffusion will be reduced in stage III (i.e. k^{III} will be increased). Therefore, interdiffusion studies on the first stage should be adequate for coating design purposes.

By extending the treatment in Crank (8) to multicomponent systems (9) it is possible to write general equations for the kinetic constants. For ternary systems they can be written for solute one as

$$k_1^I = (r_{11} + r_{12}\Delta C_2/\Delta C_1)/\sqrt{\pi} \qquad (8)$$

61

and

$$k_1^{III} = (r_{11}^{-1} + r_{12}^{-1}\Delta C_2/\Delta C_1)/\sqrt{\pi} \tag{9}$$

for each solute. The r_{1j} coefficients in (8) are from the square root diffusivity matrix, $[r]$, and the r_{1j}^{-1} coefficients in (9) are from the inverse of the square root diffusivity martix, $[r]^{-1}$.

In the binary case it is not possible to alter the kinetic constants explicitly with composition variations, but equation (8) shows that it is possible in the ternary case. In fact, k_1^I is zero in coatings with

$$\Delta C_2/\Delta C_1 = -r_{11}/r_{12} \tag{10}$$

It will be shown later that a coating with this design will eventually interdiffuse with the substrate, but its coating life is longer than that of coatings with $k_1^I > 0$.

The relationship between stage I kinetics and stage III kinetics can be determined by inspecting equations (8) and (9). Because k_1^I and k_1^{III} are linearly related to the concentration difference ratio, it follows that they are linearly related to each other. This is shown in Figure 5,

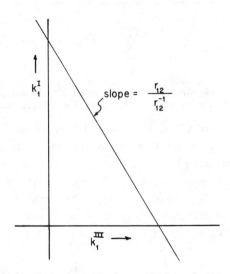

Figure 5 - The relationship between k_1^I and k_1^{III} for ternary, isomorphous systems with constant diffusivity.

a plot of k_1^I versus k_1^{III} which is a straight line. The slope of the line is equal to r_{12}/r_{12}^{-1} and in principle it is always negative. Therefore, in ternary systems the kinetic constants tend to be inversely related. In other words, any improvement in stage I kinetics (i.e. a reduction in k_1^I) through composition modification will produce a corresponding improvement in stage III kinetics (i.e. an increase in k_1^{III}). Moreover, if diffusivity data is available it is possible to calculate te kinetic constants for each stage of interdiffusion and predict a coating composition with optimum design.

Interdiffusion in Ni-Cr-Al

The use of diffusivity data in designing coatings for isomorphous systems will now be illustrated with an example that makes use of diffusivity data collected by Tu (10) for an alloy containing Ni+10at%Cr+8at%Al at 1025°C. Converting his [D] coefficients into square root diffusivity coefficients yields:

$$r_{AlAl} = 6.16 \times 10^{-6} \text{ cm (sec)}^{-1/2} \tag{11}$$
$$r_{AlCr} = 1.92 \times 10^{-6} \text{ cm (sec)}^{-1/2} \tag{12}$$
$$r_{CrAl} = 1.06 \times 10^{-6} \text{ cm (sec)}^{-1/2} \tag{13}$$
$$r_{CrCr} = 4.24 \times 10^{-6} \text{ cm (sec)}^{-1/2} \tag{14}$$

With regard to the design of interdiffusion resistant coatings it is apparent by inspection of equation (8) that when r_{12} is a positive quantity, then decreasing the amount of solute 2 in the coating will reduce k_1^I. Since in the example above r_{AlCr} is positive, it follows that decreasing the amount of Cr in the coating will decrease k_{Al}^I (i.e. it will reduce the initial rate at which Al diffuses out of the coating).

In order to illustrate the effect of Cr in more detail the coefficients of [r] were substituted into the multicomponent error function solutions and these were used to determine f_{Al} and C_{Al}^s for an example system versus time. The example system consisted of a substrate alloy containing Ni+5at%Al+10at%Cr and a coating containing Ni+8at%Al+Xat%Cr. The amount of Cr in the coating, given, by X, was varied between .04 and 15 at%.

Figures 6 and 7 give the results of the calculations. By inspection of these figures it is obvious that reducing the Cr content of the coating leads to greater resistance of Al to interdiffusion and longer life. The case when the Cr content of the coating is .04% is of special interest because it has a k_{Al}^I value of zero. As can be seen in Figure 6 the inter-

63

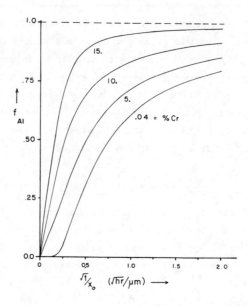

Figure 6 - The effect of % Cr in a coating on the loss of Al. The subsrate for this example contains Ni+10%Cr+5%Al while in addition to the Cr the coating contains Ni+8%Al.

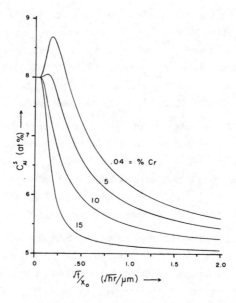

Figure 7 - The Al concentration at the surface of the coating depicted in Figure 6.

diffusion rate is zero initially, but becomes finite at a later time. A
benefit of the low initial rate is that the time for 50% loss of Al is
sixteen times longer than that for a comparable coating with 15 at%C, a
change of more than an order of magnitude.

Figure 7 gives physical insight into the mechanism by which the reduced Cr content in the coating tends to reduce Al interdiffusion. Under
these conditions a flux a Cr into the coating occurs which opposes the
flux of Al out of the coating. In fact the Al is swept back into the
coating to the outside surface as evidenced by the increase of Al concentration there in the early stages. It is only at later times when the
Cr flux has subsided that Al can leave the coating in significant amounts.

Conclusions

1. There are three stages of interdiffusion expected during the
life of a coating/substrate system. The stages can be identified by
measuring the loss of solute from the coating and the variation of surface concentration versus a time coordinate that has been normalized with
respect to coating thickness.

2. The initial and final stages of interdiffusion can be characterized
by two kinetic constants. The constants can be measured by experiment and
for selected cases they can be derived from theory.

3. When a coating and substrate are isomorphous and have constant
diffusivity, the kinetics during stage I and stage III are closely related
to the coefficients of the square root diffusivity matrix [r]. In both
binary and ternary systems, it can be shown that improvements in interdiffusion behavior during stage I result in similar improvements in stage
III. However, it is only in ternary and higher order systems that the
kinetic behavior depends explicitly on coating composition.

4. When the isomorphous model is applied to the Ni-Cr-Al system it
predicts that decreasing the Cr concentration of a coating will decrease
the interdiffusion rate of Al from the coating to the substrate. Furthermore, changes of Cr concentration on the order of 10 at% result in delaying the time to lose 50% of the Al in the coating by a factor of ten
or more.

Acknowledgement

The authors are grateful to the University of Connecticut Research
Foundation for partial support of this project and to Dr. R. H. Barkalow
for suggesting the topic.

References

1. J. E. Morral and R. H. Barkalow, _Scripta metall._, 16 (1982) 593.
2. S. R. Levine, _Met. Trans._, 9A (1978) 1237.
3. S. G. Young and G. R. Zellars, _Thin Solid Films_, 53 (1978) 241.
4. J. E. Morral, _Scripta metall._, 18 (1984) 1251.
5. M. S. Thompson and J. E. Morral, _Acta metall._, 34 (1986) 339.
6. M. S. Thompson, and J. E. Morral, _Acta metall._, 34 (1986) 2201.
7. M. S. Thompson, M.S. Thesis, The University of Connecticut, 1985.
8. J. Crank, _The Mathematics of Diffusion_ (Oxford, Clarendon Press, 1975), 14-16.
9. M. S. Thompson and J. E. Morral, unpublished research.
10. D. C. Tu, Ph.D. Disseratation, SUNY at Stony Brook, 1982.

PERFORMANCE COMPARISON OF ADVANCED AIRFOIL

COATINGS IN MARINE SERVICE

Warren D. Grossklaus, Jr.
General Electric Company
1 Neumann Way
Cincinnati, Ohio 45215

Gerald B. Katz
Naval Ships System
Engineering Station
Philadelphia, PA 19112

David J. Wortman
General Electric Company
1 Neumann Way
Cincinnati, Ohio 45215

Abstract

Several current production and developmental MCrAlY blade coatings have
been tested at sea in a "rainbow" rotor configuration on an LM2500 engine.
These coatings were developed for type 2 low temperature hot corrosion
resistance which is the primary mode of hot corrosion on LM2500 HPT blades
in U.S. Navy service. The MCrAlY coatings included: BC21 (PVD CoCrAlY),
PBC22 (vacuum plasma sprayed CoCrAlHf), BC23 (PVD CoCrAl + pack aluminide/
hafnide overcoat + Pt electroplate), PBC-23 + Cr (vacuum plasma sprayed
CoCrAlHfPt + pack chromide overcoat) and a high Cr PVD CoCrAlY coating
developed under a Navy contract. In addition, a ceramic coating (PVD yttria
(20%) stabilized zirconia) over a BC21 bond coating was also tested. An
analysis of the first stage HPT hardware after test (1077 hours) showed
that the PBC23 + Cr and BC23 coatings were clearly superior to both the BC21
and high Cr CoCrAlY coating when operating in a type 2 corrosion
environment. The ceramic coating also showed excellent corrosion
resistance. The results are explained in terms of coating microstructure
and chemistry modifications.

Introduction

The LM2500, a derivative of the CF6/TF39 aircraft gas turbine engine, is used to power several classes of U.S. Navy ships. Since 1969, General Electric and the U.S. Navy have conducted a program on board the GTS Adm. Wm. Callaghan to evaluate the effects of shipboard operating environments and power profiles on the lifetime and reliability of LM2500 components. Coating evaluations of LM2500 turbine blades and vanes from the U.S. Navy Fleet have shown that the GTS Callaghan reproduces the types and morphologies of corrosion found on Fleet engines and, when a multistage moisture separator is used, the GTS Callaghan also reproduces the corrosion rates. Previous papers on this subject (1, 2) have compared the results of several developmental and production coatings which had been tested on the GTS Callaghan.

Experimental Procedure

The U.S. Navy and General Electric have developed standardized procedures and techniques for describing hot corrosion attack of marine gas turbine components (3) which were utilized in this study. Table 1 summarizes these techniques. In this paper, testing of several recently developed coatings for improved type 2 hot corrosion resistance as well as a ceramic coating are discussed.

These coatings are:

- BC21 has been the standard for reference in the LM2500. This is a CoCrAlY coating applied by Physical Vapor Deposition (PVD) with a nominal composition of Co-22.5Cr-10.5Al-0.3Y. The coating has a uniform (nongraded) composition and a microstructure consisting primarily of two phases, -CoAl and -CoCr (Figure 1), with a third phase a fine, discrete Cr-rich phase sometimes visible.

- BC23 (Three-step) is a multi-step coating consisting of PVD CoCrAl followed by codeposition of aluminum and hafnium in a pack cementation process, and finally an electroplated platinum layer followed by diffusion through heat treatment. The coating has a nominal composition of Co-26Cr-12Al-1Hf-5Pt; however, due to the multi-step processing, a graded coating is produced with three distinct layers (Figure 2). The inner layer is essentially PVD CoCrAl, with a microstructure similar to BC21. A middle layer is produced by diffusion of Al, Hf, and Pt into the PVD CoCrAl. This layer contains a (Co, Pt) Al phase and a CoCrHf-rich phase. The outer layer is essentially a PtAl phase with some cobalt present. There is also a one-step, plasma sprayed version of this coating which is made from prealloyed powder. BC23 (either three-step or plasma) is the standard blade coating for new production engines.

 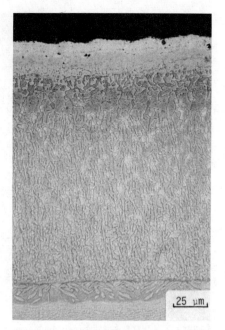

Figure 1. Microstructure of BC21 coating on an LM2500 stage 1 blade showing a uniform 2 phase microstructure.

Figure 2. Microstructure of BC23 (three-step) coating on an LM2500 stage 1 blade.

- BC22 (Plasma) is a one-step coating applied by plasma spray. Prealloyed powder is used to produce a uniform (nongraded) composition of Co-26Cr-10.5Al-2.5Hf. The microstructure of this coating (Figure 3) consists primarily of two phases, α-CoAl phase and a CoCrHf-rich phase. Vacuum plasma and argon shrouded plasma processes have been used to produce plasma BC22 coating; the plasma BC22 coating tested in this engine was applied by vacuum plasma spray.

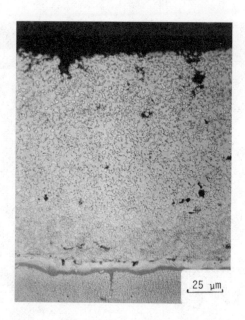

Figure 3. Microstructure of plasma BC22 coating on an LM2500 stage 1 blade
           showing a uniform microstructure.

- BC23 (Plasma) + Cr is a developmental two-step coating consisting
  of plasma sprayed BC23 and a pack cementation Cr overcoat.
  Prealloyed powder is used to produce a uniform composition of
  Co-26Cr-10.5Al-2.5Hf-5Pt (plasma BC23). A Cr pack cementation
  process is used that deposits an additive layer of α-Cr, about 0.3
  mil thick, with a diffusion zone about 0.5 mil thick (Figure 4). The
  Cr content of the diffusion zone ranges from about 40% adjacent to
  the α-Cr down to 25% at the unaffected plasma BC23 layer (Figure 5).
  The pack cementation process parameters were selected so that Al
  depletion from the plasma BC23 coating was minimized. This coating
  was developed by General Electric, as part of a NAVSEA sponsored
  Component Improvement Program, as an improved type 2 corrosion
  resistant BC23 coating.

Figure 4. Microstructure of plasma BC23 + Cr coating on an LM2500 stage 1 blade showing a single phase Cr enriched outer layer.

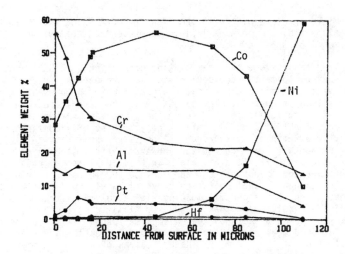

Figure 5. Composition gradient of elements in the plasma BC23 + Cr coating. The outer surface is Cr enriched. Nickel comes from the nickel base superalloy substrate.

- **NAVSEA High Cr-CoCrAlY** is a one-step coating applied by plasma spray. The coating was applied by the same vendor as GE uses for BC23 (plasma) coating. Prealloyed powder is used to produce a uniform, composition of Co-30Cr-12Al-.5Y. The microstructure of this coating (figure 6) consists primarily of two phases, a β-CoAl phase and a CoCr rich phase. This coating was developed by Pratt & Whitney Aircraft under an R&D contract from NAVSEA as an interim coating for improved type 2 corrosion resistance for marine gas turbine use. A contract goal had been a 4X improvement over BC21 in type 2 hot corrosion resistance.

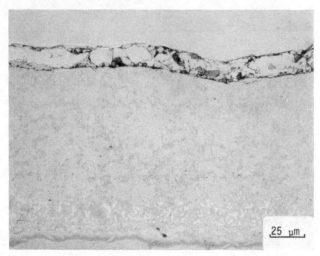

25 μm

Figure 6. Microstructure of the NAVSEA High Cr-CoCrAlY coating on an LM2500 stage 1 blade. A nickel plate is present on the surface for metallographic purposes.

- **BC21 + Ceramic** is a two-step coating consisting of a BC21 bond coat (identical to that described previously) and a thin, (0.5-2.0 mil) 20% yttria stabilized zirconia top coat applied by physical vapor deposition (figure 7). The objective of the ceramic top coat in this application is for hot corrosion protection rather than thermal barrier benefit. The limitations of yttria stabilized zirconia due to yttria leaching by vanadium (4) or yttria sulfation in high SO_3 environments (5) are recognized; however, yttria stabilized zirconia was used since it was available. In this application, vanadium contamination is not present and fuel sulfur levels were expected to be low enough so that sulfation of yttria would not occur. The PVD process produces a ceramic coating with a columnar structure. The individual columns are dense and essentially single crystals; however, porosity between the columns may permit salts to penetrate to the bond coat, resulting in hot corrosion attack and subsequent spallation of the ceramic. The objectives of testing the yttria stabilized zirconia coating were to evaluate its spallation resistance in a marine environment and to determine if salts would penetrate the porosity in the ceramic.

Results/Discussion

In previous testing, three-step BC23 provided a 1.5 - 2X corrosion life improvement over BC21 in Callaghan Service. Both coatings are limited by type 2 corrosion rather than type 1 corrosion. In this paper, test results from LM2500 engine 806/10 with the aforementioned coatings will be compared. This engine was operated for 1077 hours aboard the GTS Callaghan behind a single-stage moisture separator which results in an average sea salt ingestion level of 0.006 ppm in this installation. This is a higher salt level than is typically found on U.S. Navy Destroyers (0.002 ppm sea salt); however, this high inlet salt level has been shown to reproduce the corrosion patterns and morphologies found on destroyer engines but at a faster rate. Diesel fuel was used with an average sulfur level (shown to be a very significant factor in the type 2 corrosion rate) of about 0.4%. A power profile intended to simulate destroyer conditions was followed for part of the 1077 hours. An average of 8700 horsepower is produced by this power profile and results in high pressure turbine blade temperatures that are in the low temperature (type 2) corrosion range (1200-1400°F). Most of the remainder of the time the engine was operated at higher power, resulting in an overall average of 11,500 horsepower.

Typically blades with BC21 coating will have a coating life of 3500-4000 hours on the GTS Callaghan with this filtration system; however, borescope inspections revealed significant corrosion attack on some of the blades. During a scheduled removal of the engine, the blades were removed for careful visual examination. This examination confirmed that significant attack had occurred and, therefore, one blade pair of each group was removed for metallurgical evaluation. The remainder of the blades were returned for additional service, accumulating an additional 750 hours operation.

25 μm,

Figure 7. Microstructure of PVD BC21 + PVD yttria stabilized zirconia coating on an LM2500 stage 1 blade.

73

An LM2500 stage 1 HPT blade is shown in figure 8. Typically the areas of most severe type 2 corrosion are at the thumbprint (pressure side at 80% span and 60% chord) and root print (pressure side at 20% span and 50% chord) locations. These are the coolest locations on the pressure side of the blade due to the cooling passage design.

Figure 8. Photograph of an LM2500 stage 1 blade showing the thumbprint and rootprint locations where the most severe type 2 hot corrosion attack typically occurs.

Photographs of the stage 1 HPT blade pairs removed for evaluation are shown in figure 9. Of the metallic coatings, the PBC23+Cr was visually the best with no irregular scaling and only slight discoloration at the thumbprint of some blades. BC23 blades showed discoloration and slight scaling at the thumbprint location. Plasma BC22 had some evidence of material loss at the thumbprint location. The NAVSEA Hi Cr-CoCrAlY and BC21 coated blades had attacked regions at both the thumbprint and root print locations. The BC21 coated blades appeared to be the worst. Visually, the ceramic coated blades were in excellent condition although some spallation and other distress of the ceramic was noted, which will be discussed later.

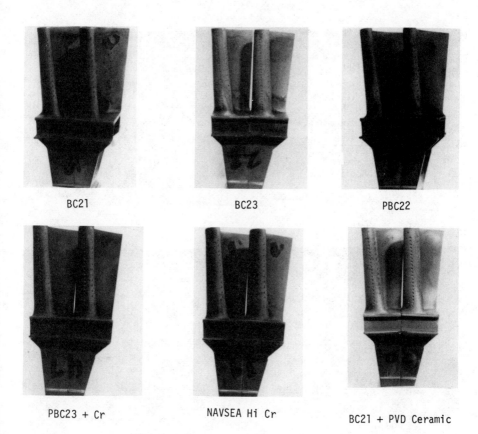

BC21   BC23   PBC22

PBC23 + Cr   NAVSEA Hi Cr   BC21 + PVD Ceramic

Figure 9.   Photographs of the stage 1 blades from LM2500 engine 806/10 after 1077 hours service on the GTS Callaghan.

The micrographs in figure 10 show the thumbprint location at 80% span which is generally the most severely attacked region of the blades. The corrosion morphology commonly referred to as Type 2 corrosion (3) was present. This morphology is characterized by a lack of alloy depletion zone beneath the scale and the presence of a dense, often layered, inner scale which is Al and Cr-rich and a Co-rich outer scale. The corrosion mechanism which produces the Type 2 corrosion in the LM2500 has been shown (6) to be due to the formation of a low melting eutectic $CoSO_4$-Na_2SO_4 salt deposit. The SO_3 content of the combustion gas is critical in the formation and stability of the $CoSO_4$. The BC21 and NAVSEA coatings are nearly penetrated and the BC23 and PBC23 + Cr coatings have only slight attack. The attack of the BC23 is confined to the outer Pt rich layer. The PBC23 + Cr coating has a thick scale which is the result of the conversion of most of the α-Cr layer to a Cr oxide. This oxide is protective, compared to the Al/Cr/Co rich oxides that typically form in type 2 corrosion. The plasma BC22 coating was intermediate in the amount of attack with typical type 2 scale present.

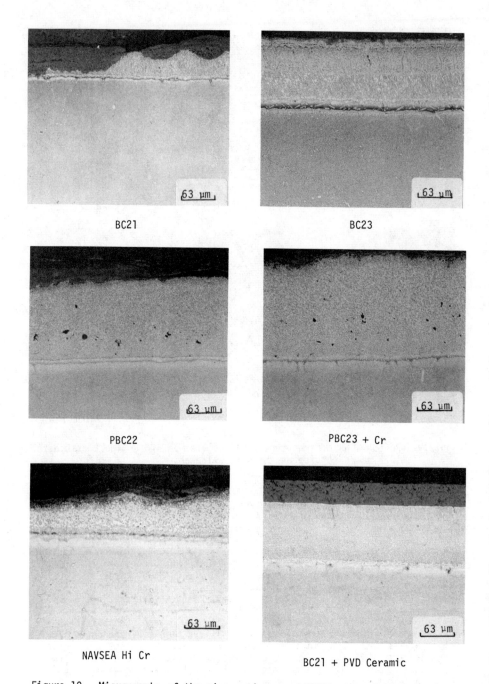

BC21

BC23

PBC22

PBC23 + Cr

NAVSEA Hi Cr

BC21 + PVD Ceramic

Figure 10. Micrographs of the six coatings on LM2500 stage 1 blades at the
thumbprint location after 1077 hours service.

The visual and optical metallographic results were used to prepare maps of the corrosion patterns and morphologies. These maps are shown in figure 11. The code used in the mapping is presented in Table 1. These maps show that the BC21 and NAVSEA Hi Cr-CoCrAlY blades had nearly all of the coating thickness consumed at the thumbprint and rootprint locations.

A more detailed evaluation of the BC21 + PVD ceramic coated blades has been performed to determine; 1) erosion needs for ceramic coatings, 2) nature of spallation and, 3) possibility of salt penetration through the ceramic.

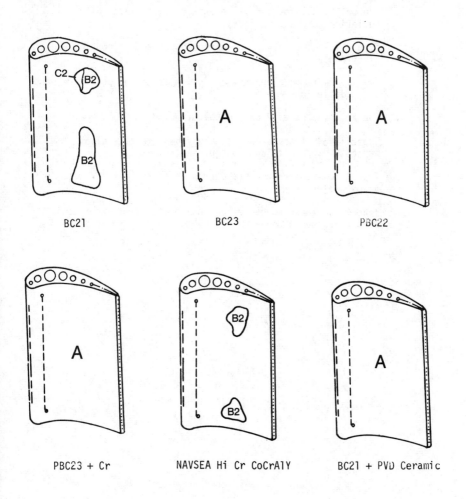

Figure 11.   Corrosion maps of the stage 1 blades for LM2500 engine 806/10 after 1077 hours service on the GTS Callaghan. The code for mapping is shown in Table 3.

TABLE 1

DOD-STD-2182/GENERAL ELECTRIC CORROSION MAPPING CODE

Type	Characteristics
A	Less than 50% loss of coating thickness
B	50-100% loss of coating thickness
1	Typical type 1 hot corrosion attack of coating - alloy depletion zone and sulfides
2	Type 2 low temperature hot corrosion attack of coating - no alloy depletion
3	Type 2 low temperature hot corrosion attack of coating - some alloy depletion
C	Coating penetrated - base metal attacked
1	Typical type 1 corrosion attack of base metal - alloy depletion zone with sulfides present
2	Type 2 low temperature hot corrosion attack of base metal - no alloy depletion
3	Type 2 low temperature hot corrosion attack of base metal - some alloy depletion

Erosion was observed at the convex side near the tip and, on some blades, has removed all of the ceramic (figure 12). This is the location on the blade where aerodynamic calculations would predict erosion by large particles to be most severe. Spallation did not appear to be the cause of loss at this location since some blades showed some remaining ceramic here as compared to other locations of the blade where spalling removed all of the ceramic (figure 13) without leaving any evidence of a thin zirconia layer. This spalling mode for PVD ceramic coatings is different from plasma sprayed TBC type zirconia coatings which leave a thin layer of zirconia after spallation (i.e. the spalling occurs within the zirconia rather than at the zirconia - bond coat interface). In spite of the fact that spallation occurred without leaving any zirconia, no evidence of hot corrosion attack of the BC21 bond coat could be found. Optical metallography showed no evidence of sulfides in the thin β- CoAl depleted layer at the surface of the BC21 and the amount of alloy depletion was only 0.2 mil which is similar to the amount of alloy depletion on the as-coated blades (figure 14). Electron microprobe analysis could not detect any sulfur at the bond coat - zirconia interface or within the bond coat (figure 15).

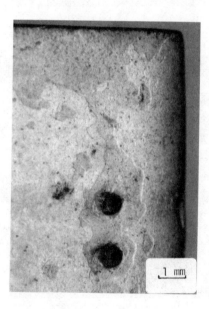

Figure 12.   Photograph of the convex side near the tip of a BC21 + PVD ceramic coated blade showing evidence of erosion and spallation.

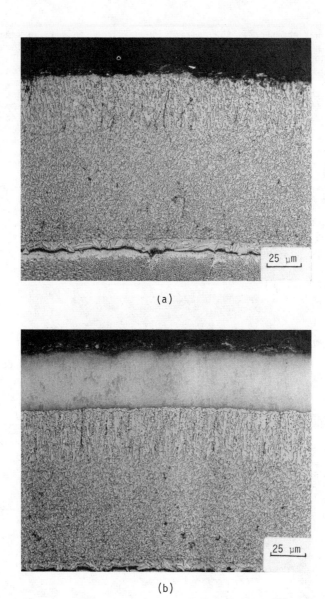

(a)

(b)

Figure 13.   Micrograph of the leading edge of a BC21 + PVD ceramic coated
blade showing adjacent spalled (a) and non-spalled (b) areas.
No evidence of hot corrosion attack is present at either location.

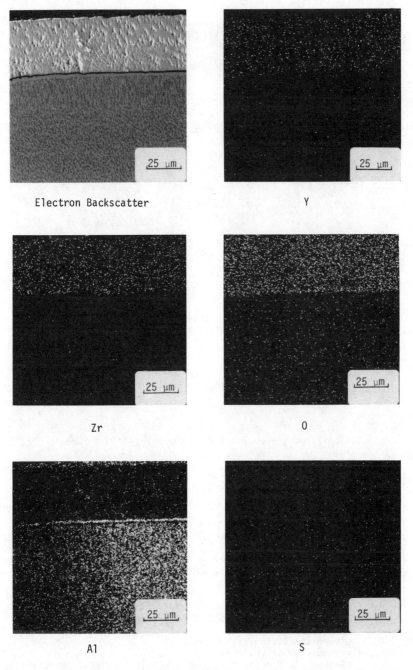

Figure 14. Electron microprobe analysis of the BC21 + PVD ceramic coating showing no evidence of sulfur in the BC21 depletion zone at the BC21/ceramic interface.

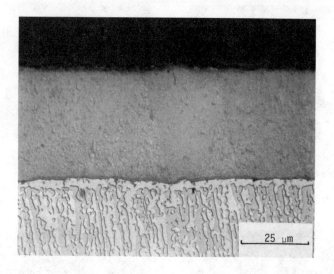

Figure 15. Micrograph of the BC21 + PVD ceramic coating at the pressure side of the blade showing the thin depletion zone.

Conclusions

The conclusions from the evaluation of the metallic and ceramic coatings are;

1. A BC23 (plasma) + Cr coating provides improved Type 2 corrosion resistance due to the Cr enrichment of the outer layer of the coating.

2. BC23 (three step) provided type 2 hot corrosion resistance greater than 4X BC21 which is considerably better than previous experience.

3. The Hi Cr CoCrAlY did not provide the 4X increase in type 2 corrosion resistance over BC21 which had been predicted based on burner rig tests.

4. PVD ceramic coatings show potential for use as corrosion resistant coatings, however several areas of development are required;

   a. improved spallation resistance,
   b. improved erosion resistance,
   c. further evaluation of needs for prevention of salt penetration.

References:

1. J.J. Grisik, R.G. Miner, and D.J. Wortman, "Performance of Second Generation Airfoil Coatings in Marine Service," Thin Solid Films, 73 (1980), 397-406.

2. D.J. Wortman, "Performance Comparison of Plasma Spray and Physical Vapor Deposition BC23 Coatings in the LM2500," J. Vac. Sci. Technol. A, 3, (6), (1985), 2532-2536.

3. DOD-STD-2182, "Standard Procedure for the Metallurgical Examination of Marine Gas Turbine Hot Section Components."

4. P.A. Siemers and D.W. McKee, U.S. Patents 4,328,285, May 4, 1982.

5. R.L. Jones, C.E. Williams and S.R. Jones, "Reaction of Vanadium Components and Ceramic Oxides," (NRL Memorandum Report 5603, July 2, 1985).

6. K.L. Luthra and D.A. Shores, J. Electrochemical Soc., 127 (10) (1980), 2202.

COATING/SUBSTRATE INTERACTIONS AT HIGH TEMPERATURE

K.L. Luthra and M. R. Jackson

General Electric Company
Corporate Research & Development
Schenectady, New York 12301

Abstract

Metal surface temperatures in future aircraft engines may exceed 1220°C (2330°F), to take advantage of single crystal capability. Oxidation at these temperatures requires use of protective coatings. Overlay coatings are thicker environmental barriers than are the diffusion aluminides, but considerable interdiffusion with the substrate can occur at elevated temperatures. To evaluate interaction severity, diffusion couples were made of single crystal superalloy samples (similar to advanced alloy chemistries) coupled to chill-cast MCrAl(Y) samples. The simulated overlay compositions ranged from 18.7-30.9 a/o Cr, 9.6-23.2 a/o Al, 0-0.2 a/o Y in the Ni, Co and Fe-base MCrAl(Y) alloys. Couples were exposed 50-200h at 1150°C and 1225°C. Electron microprobe chemical analysis and metallographic characterization of the interdiffusion zone was performed on each couple. The effects of the MCrAl(Y) chemistry on interdiffusion with the superalloy are discussed.

Introduction

Metal temperatures on jet engine turbine blades currently flying have bulk values in the 1000-1050°C (1830-1920°F) range. In localized areas on the airfoil surface, especially during transient conditions, temperatures may be much greater, in the 1100-1150°C (2000-2100°F) range. Bulk temperatures are expected to increase over the next decade, as designers take advantage of the improved performance of Ni-base single crystal alloys presently in development or scale-up, to achieve greater thrust and fuel efficiency (1,2). As the bulk metal temperatures are increased, localized surface temperatures may approach the 1230-1260°C (2250-2300°F) range.

These higher temperatures would increase the substrate/coatings interdiffusion. The interdiffusion distance is approximately proportional to D. Thus, using a normal activation energy for solid-state diffusion of 60-100 K cal/mole (3), an increase in temperature from a present maximum of 1150°C to a future maximum of 1250°C would increase interdiffusion distance by a factor of 2 to 3. This increased interdiffusion may require that thicker coatings be used to maintain an adequate reservoir of the protective elements. More importantly, the underlying superalloy may be chemically altered to a significantly greater depth. The delicate balance of γ and γ' chemistries resulting in the optimum superalloy properties will certainly be perturbed; furthermore, the chemistry in the interdiffusion zone may move in the direction of precipitation and growth of additional undesirable phases (4). These changes could have a detrimental effect on the mechanical properties of the substrate/coating system, and diffusion barrier layers may have to be used (5).

Substantial interdiffusion of coating and substrate can lead to markedly different oxidation behavior (6,7). In Figure 1, oxidation in a one-hour cyclic mode between room temperature and 1225°C is compared for a

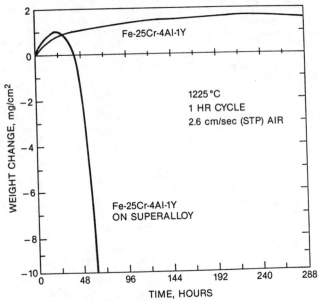

Figure 1   A comparison of the oxidation behavior of a bulk plasma sprayed Fe-25Cr-4Al-1Y coupon and a 0.013 cm coating of the same composition on 0.5cm single crystal superalloy pin in cyclic tests at 1225°C.

FeCrAlY coupon (taken from a thick deposit produced by low pressure plasma deposition - LPPD), and a superalloy pin LPPD coated with the same FeCrAlY. Severe oxidative attack occurred on the pin from the chemical changes of the interdiffused materials, compared to the excellent oxidation resistance of the flat specimen made solely of FeCrAlY.

A number of studies have considered substrate/coating interaction (7-11), both with and without the influence of an oxidizing environment. These studies have concentrated on interactions in simple ternary and quaternary systems. The present study treats the case of a complex chemistry superalloy and its interactions with Co-, Fe- and Ni-CrAl(Y) chemistries. Oxidation effects have been avoided, and the intent was to use samples thick enough that the couples were semi-infinite relative to the diffusion times and temperatures. This condition was achieved for studies at 1150°C, but at 1225°C, the complete specimen thicknesses showed altered chemistries.

Experimental Procedures

Interdiffusion studies were conducted on diffusion couples bonded together by heating in compression. The nominal compositions of alloys used in this study are listed in Table I. All of these alloys, except the superalloy used to simulate a substrate, were chill-cast to dimensions of 1.6cm x 2.5cm x 1.6cm and subsequently machined to samples of 1.3cm x 1.3cm x 0.1 - 0.15 cm. The superalloy samples were machined from single crystal castings. The samples were polished on SiC papers from 240 grit down to 600 grit, cleaned with xylene and acetone, and subsequently bonded together by heating in compression at 1100°C and 65 atm for 30-40 minutes. The diffusion couples involved conventional MCrAl(Y) coatings (M = Ni,Co,Fe) and single crystal samples of the superalloy. The diffusion direction for the latter alloy was along an axis normal to the [001] direction; no attempt was made to exactly determine this axis.

The diffusion couples were exposed in a tube furnace in a gettered argon environment for 50-200 hours at 1150°C and 1225°C. Unless otherwise noted, the samples were slowly cooled in stages to 100°C in about 15 minutes. In some cases, they were rapidly cooled to about 400°C in one minute and then slowly cooled to room temperature. Subsequently, the samples were sectioned normal to the couple interface, and the extent of interdiffusion determined by microprobe analysis. The interdiffusion results so obtained were substantiated by metallographic characterization of the interdiffusion zone. Since the primary purpose of this study was to understand the bulk diffusion processes, detailed investigations of the phases formed in the diffusion zone were not conducted.

Table I. Nominal Compositions of Alloys
atomic percent (weight percent)

	Ni	Co	Fe	Cr	Al	Other
	Bal	-	-	30.8(30)	9.9(5)	-
	Bal	-	-	19.3(20)	23.2(12.5)	.2Y(.3)
MCrAl(Y)	-	Bal	-	30.9(30)	10.0(5)	-
	-	Bal	-	21.3(22)	22.4(12)	-
	-	-	Bal	30.0(30)	9.6(5)	-
	-	-	Bal	18.7(20)	22.6(12.5)	.2Y(.3)
Superalloy	Bal	7.5(7.5)	-	10.5(9.2)	8.1(3.7)	5.2Ti(4.2), 1.9W(6.0), 1.3Ta(4.0), 0.93Mo(1.5), 0.32Nb(0.5)

Results and Discussion

The interdiffusion can be related to the governing phase equilibria by measuring the compositional variations and plotting those variations onto the appropriate phase diagram (12-14). Before presenting the results of this study, we will show the changes that can occur in MCrAl alloys. Figure 2a shows schematically the changes in Al and Cr contents resulting from oxidation of a pin of a NiCrAl alloy. Over most of the pin cross-section, the concentrations are unchanged from before oxidation. Near the surface, however, both Cr and Al have been depleted. On the ternary NiCrAl diagram, the compositions are plotted schematically. The point A represents the original Cr and Al values that exist over most of the pin after oxidation. The region over which the composition has been altered is represented by the line from A to B. It must be noted that the line AB is not a conventional diffusion path. Diffusion paths are generated from semi-infinite diffusion couples. Whether diffusion is studied after one hour or 1000 hours, a diffusion path is time-invariant (13). For the finite system of an oxidizing pin, the compositional path AB is not time-invariant. Since Al and Cr (and Ni) are lost from the system by oxidation, the B terminal of the path will vary in location, depending on time of oxidation.

In Figure 2b, the changes in Al and Cr concentrations resulting from oxidation of a NiCrAl pin coated with another NiCrAl composition, higher in Cr and Al, are shown schematically. Over most of the substrate, the concentrations are as they originally were before exposure, as given by point A' in the ternary diagram. Near the coating/substrate interface, interdiffusion has altered substrate composition from A' and coating composition from A. At the outer surface, oxidation has caused a depletion of Al and Cr much as in Figure 2a and the new composition is represented by B. For the example shown, interdiffusion and oxidation have so depleted the coating that no region of the coating remains at composition A. This indicates that the composition path is definitely not time-invariant. Depending on coating thickness, the profile will depart from composition A quickly for thin coatings, and more slowly for thicker coatings. It is apparent that as the chemistry of the coating moves from point A to a composition more like point A' (the substrate chemistry), the oxidation resistance of the coated pin will be more like that of an uncoated pin. Homogenization of a too-thin coating with the substrate naturally leads to rapid oxidation as the surface concentrations of Al and Cr fall to unprotective levels.

For the case of a NiCrAl substrate coated with a FeCrAl material, a ternary representation is no longer possible and the compositional path cannot be mapped simply in two dimensions. Such a system is shown schematically in Figure 2c. For this case, the compositional profile shows variations in Fe content as well as for Cr and Al. If the relative interdiffusion of Fe and Ni is small, the compositional path may be plottable as paths in the NiCrAl and FeCrAl ternary regions with a dashed line joining the two paths at the FeCrAl/NiCrAl interface. If substantial Fe and Ni interdiffusion have occurred however, a true quaternary section (Figure 2d) is needed to allow a three-dimensional representation of the compositional path.

The composition variations discussed above depend (obviously) on the composition of the substrate alloy. A consideration of superalloy chemistry is needed to differentiate between advanced single crystal alloys and the traditional equiaxed blade alloys such as IN-738 and Rene'80. The equiaxed alloys are generally high in Cr content (14-16a/o) and low in Al content (6-8a/o), and contain substantial amounts of C, B and Zr for maintenance of grain boundary strength and ductility (15). These alloys generally are protected with aluminide (NiAl) coatings or overlay MCrAl(Y) coatings high in Cr (25-30 a/o) and of intermediate Al content (10-25 a/o). Compared to cast blade alloys, the single crystal alloys being developed for near-term future engines are generally lower in Cr content (8-11 a/o)

88

and higher in Al content (8-11 a/o), with little or no C, B, or Zr (1,2). They typically have substantial additions of γ' strengtheners Ti, Ta and Nb. These elements apparently substitute on the aluminum sublattice in the γ' (Ni$_3$Al). In the advanced single crystal alloys, the sum of (Al + Ti + Ta + Nb) that contributes to γ' formation is in the range of 14-16 a/o, compared to a value typically in the range of 10-14 a/o for equiaxed compositions, adjusted for the depletion of that sum by carbide formation.

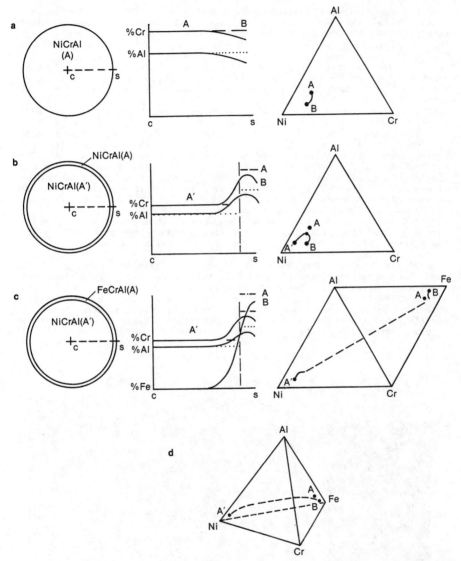

Figure 2   Schematic illustration of oxidation and interdiffusion effects on chemistry for a) a Ni-Cr-Al alloy; b) a Ni-Cr-Al alloy coated with Ni-Cr-Al; c) a Ni-Cr-Al alloy coated with Fe-Cr-Al, where the dashed line indicates the portion of the path describable only in three dimensions, as in d.

89

The alloy chosen for representing advanced blade alloys is in line with the general direction described above. The Cr and Al contents of 10.5 and 8.1 a/o, respectively, are in the ranges noted for advanced single crystals. The sum of (Al+Ti+Ta+Nb) is approximately 15 a/o, which is expected to promote a large volume fraction of γ'. The alloy contains small amounts of γ solid solution elements, W and Mo, as well as 7.5 a/oCo, but it contains no C, B, or Zr.

In the following presentation, results of diffusion studies are presented for the superalloy and either a Ni, Co, or Fe base MCrAl(Y) material. For each MCrAl(Y), two compositions are considered: one of high Cr, low Al content (approximately 30 a/o Cr and 10 a/o Al or 30 w/o Cr and 5 w/o Al), and a second of low Cr, high Al content (approximately 20 a/o Cr, 22.5 a/o Al or 20 w/o Cr and 12.5 w/o Al). The Y contents might be expected to influence the oxidation behavior of these materials, but the 0-0.2 a/o Y additions will play an incidental role in establishing the interaction paths of the coatings with the superalloy substrate. If not specified, compositions are in weight percent in the following discussions.

Superalloy - NiCrAl(Y) Couples. Figure 3 shows the microstructure developed in 50 hours at 1150°C. Structures were very similar after 200 hours, except that diffusion had occurred over a greater distance. The superalloy/high Cr, low Al (30 w/o Cr, 5 w/o Al) couple showed an extremely flat Al concentration profile, since both the alloys have nearly equivalent Al contents (8.1 vs 9.9 a/o). A larger driving force exists for both Cr and Co, and Figure 4 shows diffusion distances of 300-400 μm. Figure 5 shows the profiles for Ti and Ta after 200h at 1150°C. Penetration of both elements into the low Cr, high Al was to a greater distance than into the high Cr, low Al terminal. Behavior of W, Mo and Nb was similar to Ta.

Figure 3 shows that the high Cr, low Al terminal coupled to the superalloy shows considerable Kirkendall void formation. This is caused by the faster outward diffusion of Cr from the high Cr, low Al alloy than the inward diffusion of Co, Ti, Ta, W, Mo and Nb. In addition, a clear band of structure about 35 μm in width has developed at the interface. From the profile, this is single phase γ, with composition varying smoothly from the γ phase composition of the superalloy to that of the single phase high Cr,

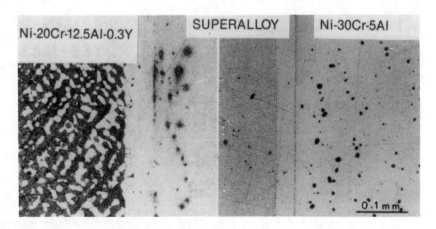

Figure 3   Optical micrographs of a Ni-30Cr-5Al/Superalloy/Ni-20Cr-12.5Al-0.3Y couple exposed for 50 hours at 1150°C. Superalloy is 7.5 Co, 9.2 Cr, 3.7 Al, 6 W, 4 Ta, 4.2 Ti, 1.5 Mo, .5 Nb.

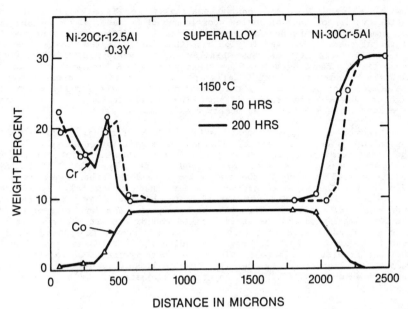

Figure 4   Results of microprobe chemical analysis for interdiffusion of
Cr and Co in Ni-20Cr-12.5Al-.3Y/superalloy/Ni-30Cr-5Al couples
exposed at 1150°C. Superalloy is 7.5 Co, 9.2 Cr, 3.7 Al, 6 W,
4 Ta, 4.2 Ti, 1.5 Mo, .5 Nb.

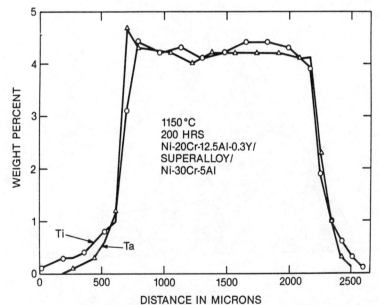

Figure 5   Results of microprobe chemical analysis for interdiffusion of
Ta and Ti in Ni-20Cr-12.5Al-0.3Y/superalloy/Ni-30Cr-5Al couples exposed for 200 hours at 1150°C. Superalloy is 7.5 Co,
9.2 Cr, 3.7 Al, 6 W, 4 Ta, 4.2 Ti, 1.5 Mo, .5 Nb.

low Al alloy. This is seen schematically in Figure 6. An accurate representation of this diffusion profile cannot be made, since the superalloy composition (S) contains many elements in addition to Ni, Cr and Al. If the Al content was used to plot onto the NiCrAl diagram (16-19), the 8.1 a/o level of the high Cr, low Al alloy would clearly put it in the single phase field. Plotting point S at the approximate Al content equal to the Al+Ti+Ta+Nb sum, the point S falls in the γ + γ' field, corresponding to the actual microstructure. The interaction between the superalloy and the high Cr, low Al alloy (point 1) shows the profile leaving the γ + γ' field to a single phase γ field, as observed experimentally.

As seen in Figure 3, the low Cr, high Al terminal coupled to the superalloy is quite different. Whereas the superalloy consists of γ + γ', the other terminal is two phase γ + β (NiAl). At the joint between the two, continuous planar layers are formed. After 50h at 1150°C, the layer on the superalloy side is about 40μm thick, and the other layer is about 12μm thick. The layer formations are described schematically in Figure 6 by the path S<-->2. The addition of Al to the superalloy surface converts that region of the γ + γ' structure to solely γ'. The departure of Al from the terminal 2 side and the arrival of Co and Ni convert the γ + β structure at the interface to solely γ. Since no lateral two-phase growth is seen in Figure 3 and the interfaces are planar, the paths are dashed in the two-phase fields of Figure 6. The dashed path indicates an interface with compositions governed by equilibrium tie-lines. Two-phase growth is represented by a solid curve which crosses tie-lines.

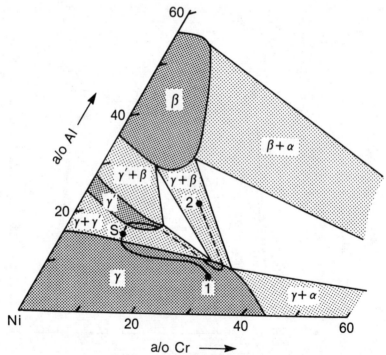

Figure 6   Schematic illustration of the compositional profiles for Ni-30Cr-5Al/superalloy (1->S) and Ni-20Cr-12.5Al-0.3Y/superalloy (2→S) couples represented on the Ni-Cr-Al ternary phase diagram - 1150°C isothermal section.

Superalloy - CoCrAl Couples. The micrographs of the couples exposed for 50 hours at 1150°C and 1225°C are shown in Figures 7 and 8, respectively. The results of the microprobe analysis for Ni, Co, Cr, and Al for the sample in Figure 8 are shown in Figure 9. Although both of the CoCrAl alloys used here have essentially two-phase $\gamma+\beta$(CoAl) structures, their results are somewhat different.

Let us first discuss the results for the high Cr, low Al side. A region 100μm wide of the CoCrAl has been depleted of the β phase at 1150°C. Growth of platelets of β(Ni,Co)Al has occurred to a depth of about 35μm into the superalloy. Exposure at 1225°C (Figure 8) produced even greater recession of β in the CoCrAl, approximately 300μm, but did not result in β formation within the superalloy. The microprobe analysis results (Figure 9) show that Ni and Co diffusion are essentially counterbalancing. However, the Cr diffusion from the CoCrAl exceeds Al (and Ti, Ta, Mo, W, Nb) from the superalloy so that Kirkendall voids form in the CoCrAl. The total diffusion distance for the major elements is seen to be in the range of 350μm. The absence of β(Ni, Co)Al in the superalloy at 1225°C suggests that the solubility of aluminum in the Ni, Co, Cr, Al γ solid solution is greater at 1225°C than at 1150°C.

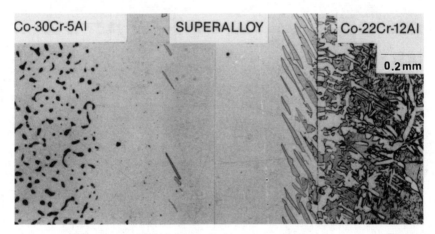

Figure 7   Optical micrographs of a cross-section of a Co-30Cr-5Al/superalloy/Co-22Cr-12Al couple exposed for 50 hours at 1150°C. Superalloy is 7.5 Co, 9.2 Cr, 3.7 Al, 6 W, 4 Ta, 4.2 Ti, 1.5 Mo, .5 Nb.

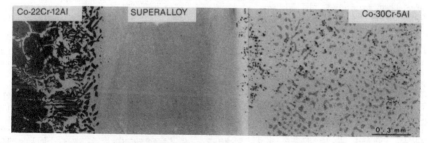

Figure 8   Optical micrographs of a cross-section of a Co-30Cr-5Al/superalloy/Co-22Cr-12Al couple exposed for 50 hours at 1225°C. Superalloy is 7.5 Co, 9.2 Cr, 3.7 Al, 6 W, 4 Ta, 4.2 Ti, 1.5 Mo, .5 Nb.

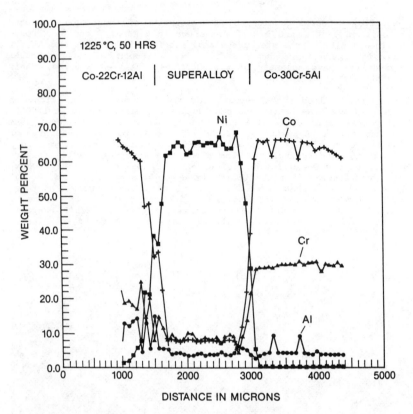

Figure 9   Results of microprobe chemical analyses for interdiffusion in
a Co-30Cr-5Al/superalloy/Co-22Cr-12Al couple exposed for 50
hours at 1225°C. Superalloy is 7.5 Co, 9.2 Cr, 3.7 Al, 6 W, 4
Ta, 4.2 Ti, 1.5 Mo, .5 Nb.

The phase relations for these results are presented in Fig. 10. As Cr
diffuses into the $\gamma + \gamma'$ superalloy at 1150°C, the interfacial region composition varies from the $\gamma + \gamma'$ field towards the $\gamma + \beta$ field, and precipitation of a very small amount of β occurs within the superalloy. Diffusion
of Cr at 1225°C moves the interfacial composition in the same direction,
but the expanded solubility γ phase is formed without β precipitation.
Note that the Co-Al binary system forms no γ' phase (19). Therefore, the
diffusion of Co into the superalloy tends to stabilize $\gamma + \beta$ over $\gamma + \gamma'$.

In comparison to the high Cr, low Al alloy, greater microstructural
changes are observed for diffusion with the low Cr, high Al alloy (see Figures 7 and 8). Much greater precipitation of β is seen in the superalloy
at 1150°C, to a depth of 65μm, and β is now present in the superalloy at
1225°C to a depth of 100μm. Decreases in the amount of β present in the
CoCrAl occur to depths of 80 and 300μm at 1150 and 1225°C, respectively.
The simultaneous diffusion of Cr, Al and Co from the low Cr, high Al CoCrAl
into the superalloy pushes the interfacial composition from the $\gamma + \gamma'$
field toward the $\gamma + \beta$ field at both the temperatures (Figure 10). The
intersection with the two-phase tie lines occurs at a greater Al content,
and therefore at a greater volume fraction of β for the low Cr, high Al
CoCrAl than for the high Cr, low Al CoCrAl.

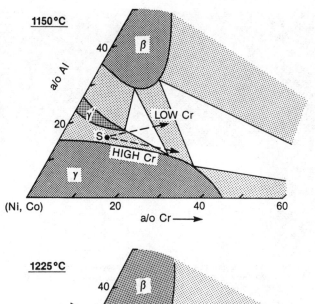

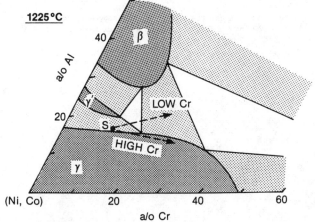

Figure 10   Schematic illustration of the direction of compositional alteration for superalloy/CoCrAl couples exposed at 1150°C or 1225°C, represented on isothermal sections through the Ni-Co-Cr-Al quaternary phase diagram.

Superalloy - FeCrAl(Y) Couples. The most dramatic interdiffusion was observed for the superalloy - FeCrAl(Y) couples. The micrographs of a couple exposed for 50 h at 1225°C are shown in Fig. 11. The corresponding reaction paths and the results of microprobe analysis are shown in Figures 12 and 13, respectively. Similar structures and profiles were developed at 1150°C after 50 hours (Table II).

Let us first consider the results for the high Cr, low alloy. Due to diffusion of Ni and Co from the superalloy, the high Cr, low Al FeCrAl was transformed from α single phase (bcc Fe solid solution) to an $\alpha + \gamma$ mixture near the interface. Within the superalloy, diffusion has resulted in a degradation from the original single crystal form. The large differences in fcc lattice parameter for Ni and Fe (20) may be the cause. Newly formed grain boundaries serve as short-circuit paths for deep penetration. Along these boundaries new phases are formed and grow laterally into the $\gamma + \gamma'$ structure (Figure 11). Although detailed analyses have not been made, a β

95

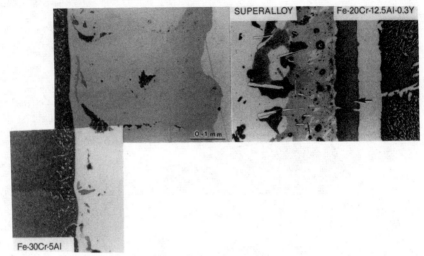

Figure 11   Optical micrographs of a cross-section of a Fe-30Cr-5Al/superalloy/Fe-20Cr-12.5Al-0.3Y couple exposed for 50 hours at 1225°C. Superalloy is 7.5 Co, 9.2 Cr, 3.7 Al, 6 W, 4 Ta, 4.2 Ti, 1.5 Mo, .5 Nb.

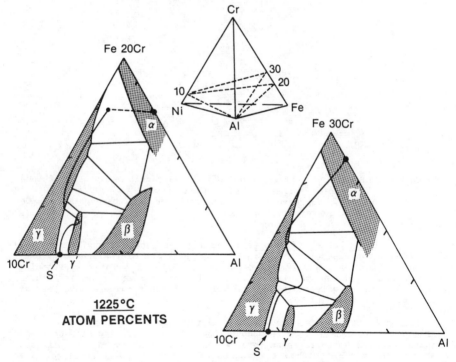

Figure 12   Schematic illustration of the compositional profiles of Fe-30Cr-5Al/superalloy/Fe-20Cr-12.5Al-0.3Y couples represented on 1225°C sections through the Fe-Cr-Al-Ni quaternary phase diagram.

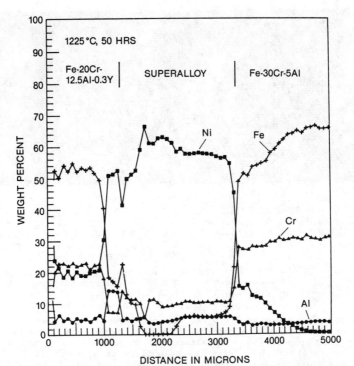

Figure 13   Results of microprobe chemical analyses for interdiffusion in
             a Fe-30Cr-5Al/superalloy/Fe-20Cr-12.5Al-0.3Y exposed for 50
             hours at 1225°C. Superalloy is 7.5 Co, 9.2 Cr, 3.7 Al, 6 W, 4
             Ta, 4.2 Ti, 1.5 Mo, .5 Nb.

(Ni,Fe)Al precipitate is likely: the addition of Fe to γ reduces maximum
Al solubility (19-21). Note that Al is an α stabilizer in Fe and has less
than 2 a/o solubility in γ Fe. As Fe diffuses along grain boundaries, the
$\gamma + \gamma'$ regions may be converted to $\gamma + \beta$, as shown schematically in Figure
12.   The results of microprobe analysis (Figure 13) indicate a peak in Al
content and a minimum in Cr content in the regions of interfaces between
the superalloy and each FeCrAl(Y) alloy. This result was duplicated at
1150°C.

     The interface between the superalloy and the low Cr, high Al FeCrAlY
alloy was even more complex.   The α FeCrAlY was transformed into a $\gamma + \alpha$
mixture for at least 200μm, due to the arrival of Ni and Co from the
superalloy. A 60μm layer of γ was adjacent to the FeCrAlY, followed by a
60μm layer of γ'. Between the γ' layer and the original superalloy composition was another 250μm of complex structure.

     A more detailed view of the interfacial region between the low Cr,
high Al alloy and the superalloy is shown in Fig. 14, along with the
results of microprobe analysis for different phases. The γ phase in the
αFeCrAlY matrix contains only 4.5 w/o Al, compared to 12.5 w/o Al in the
parent FeCrAlY.   Considerable amounts of Al have   diffused into the
superalloy to form the continuous γ' layer (13.5 w/o Al) and the large
pockets of γ' in the complex region. The Cr content of the γ' phase is in
the range of 7-8 w/o. The Fe content of the γ' varies from 18 w/o against
the γ layer to ~ 13 w/o in the region near the unaltered $\gamma + \gamma'$ structures.
Many new precipitates in the form of platelets have formed in the superalloy. The sum of Fe, Ni, Cr, and Al in these phases amounts to 3--33.5 w/o,
with Al present at ~0.5 w/o. The remainder consists of W, Mo, Ti, Ta, and

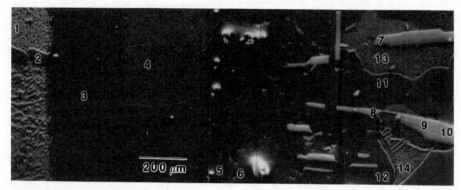

	1	2	3	4	5	6	7	8	9	10	11	12	13	14
Fe	45.1	50.7	46.0	18.0	28.2	17.0	9.0	9.0	8.0	8.3	21.6	22.3	12.5	13.2
Ni	22.3	22.0	25.1	49.1	37.4	49.5	12.2	12.0	12.2	11.5	42.4	41.4	52.6	49.9
Cr	19.5	19.0	18.4	7.6	15.4	7.9	11.8	11.9	11.2	10.8	13.7	13.4	7.0	8.6
Al	6.0	4.5	4.4	13.6	4.5	13.7	0.5	0.5	0.5	0.4	4.9	4.7	13.1	12.3
	γ+α	γ	γ	γ'	γ	γ'	TCP	TCP	TCP	TCP	γ	γ	γ'	γ'

Figure 14   Details of microstructure and phase chemistries (in w/o) for the 50 hour - 1225°C couple between the superalloy (at right) and Fe-20-Cr-12.5Al-0.3Y (at left). Superalloy is 7.5 Co, 9.2 Cr, 3.7 Al, 6 W, 4 Ta, 4.2 Ti, 1.5 Mo, .5 Nb.

Nb. No detailed phase analysis has been performed, but these compositions correspond to those of the topologically close packed (TCP) phases, and are probably σ or χphases (4). The elements W, Mo and Ti are strong α stabilizers, and Ta and Nb have very low solubility in γFe as well (19,22). As Fe diffuses into the γ + γ' structure of the parent superalloy, those elements will be rejected away from the incoming Fe. Since these elements are relatively slow diffusers, they become concentrated in localized areas and eventually cause precipitation of a TCP phase. The propensity for σ formation is known to be high in fcc structures with high Fe concentration (23).

The major portion of the reaction path can be described by the path shown in Fig. 12b (24,25). Starting from the α terminal and crossing to

Table II - Comparison of Microstructural Changes

Superalloy/MCrAl(Y) Couples After 1150°C/50h

	Microstructural Alteration	
	Superalloy Side (γ+γ' initially)	MCrAl(Y) Side
NiCrAl(Y)		
γ high Cr/low Al	γ layer~35μm	--
γ+β low Cr/high Al	γ' layer~40μm	γ layer~12μm
CoCrAl		
γ+β high Cr/low Al	γ+β layer~35μm	γ layer~100μm
γ+β low Cr/high Al	γ+β layer~65μm	{reducedβ~80μm {γ+β layer
FeCrAl(Y)		
α high Cr/low Al	complex ~ 350μm*	α+γ layer~60μm
α low Cr/high al	{γ'~50μm {complex~225μm*	{α+γ layer>185μm {γ layer~55μm

* no longer single crystal in altered region

98

the superalloy, the α is transformed to $\alpha + \gamma$, then a layer of γ, a layer of γ' and a region of predominantly $\gamma + \gamma'$ are formed. At the end of that path, TCP phases from in the superalloy.

To investigate whether the structure in Figure 11 formed during diffusion or during cooling, the couple was heated to 1225°C and was then cooled to 400°C within one minute. Much of the γ structure in the FeCrAl terminal was dissolved into the α phase by this treatment, and the fine γ' (in the superalloy beyond the coarse γ' and TCP phases) was solutioned for a distance of 200μm further into the superalloy. However, the layered structures of γ and γ' as well as the $\gamma + \gamma'$ + TCP region were unchanged.

These results show that the composition of ~125μm FeCrAlY coatings, such as the one used in Figure 1, would change quite rapidly at 1225°C. The presence of the low Al γ phase in the FeCrAl(Y) interdiffused coatings is probably responsible for the poor oxidation behavior noted in Figure 1. This, coupled with the extreme depths of perturbed microstructure in the superalloy, makes FeCrAlY compositions impractical as superalloy coatings at this temperature.

Summary and Implications

Table II summarizes the microstructural changes occurring during interdiffusion of the superalloy and MCrAl(Y) couples after 50 h at 1150°C. The structures which formed increase in complexity from Ni- to Co- to FeCrAl couples. The dimensions of the regions showing microstructural change also increase in the same order. The total region over which compositional changes occur is large for all three systems, 300- 500μm.

For the NiCrAl(Y) systems, no microstructural change was noted in the high Cr material, while a thin γ layer developed in the low Cr material. The superalloy developed a 35-40μm layer of either γ or γ'.

For the CoCrAl systems, a thick layer of β-depleted structure formed, with no β remaining in that layer for the high Cr material. The superalloy developed precipitation of β to a depth of 35-65μm.

For the FeCrAl(Y) systems, the high Cr α structure showed a thick layer of $\gamma + \alpha$ forming, while the low Cr alloy developed a very thick $\gamma + \alpha$ region and a sizable layer of γ. On the superalloy side, the single crystal structure was replaced with multiple grains in the very extensive interdiffusion zone. On the superalloy side in contact with the low Cr alloy, this zone was a coarse microstructure of γ, γ', and a phase rich in the refractory elements, and probably a TCP phase such as σ or χ. On the superalloy side in contact with the high Cr alloy, a mixed structure of γ, γ' and β was formed.

The exposure of 50 hours at 1150°C is not an extreme condition for what is anticipated to be the future need for coatings performance. The results indicate that coatings of 100μm thickness may suffer radical chemical alteration in very short periods of exposure. The change in chemistry due to oxidation will be an additional factor. The most effective coatings, in terms of microstructural stability and oxidation resistance, will remain the NiCrAlY system alloys, modified with further additions to achieve chemical near-equilibrium.

REFERENCES

1. F.L. VerSnyder and E.R. Thompson, Alloys for the Eighties (Greenwhich, CT: Climax Molybdenum Co., 1982) 69-84.

2. M.Gell, D.N. Duhl, and A.F. Giamei, Superalloys 1980 (Metals Park, OH: Am. Soc. Metals, 1980) 205-214.

3. P.G. Shewmon, _Diffusion in Solids_ (New York, NY: McGraw Hill, 1963) 66.

4. C.T. Sims, _The Superalloys_ (New York, NY: John Wiley, 1972) 259-284.

5. K.L. Luthra, J.Vac. Sci. and Tech. A., 3 (1985), 2574-2577.

6. M.R.Jackson, J.R. Rairden and L.V. Hampton, "Coatings for Directional Eutectics," (General Electric Corporate Research & Development Report 74CRD187, 1974).

7. K. Luthra and M.R. Jackson, (General Electric Company: unpublished research, 1985).

8. S.R. Levine, Met. Trans. A., 9A (1978), 1237.

9. J.A. Nesbitt (Ph.D. thesis, Michigan Tech. Univ., 1984, available as NASA TM-83738).

10. R.J. Hecht, R.H. Barkalow, S.D. Houser, and R.L. Shamakian (Report F33615-78-C-5206, United Technology Corp., 1980).

11. M.R. Jackson and J.R. Rairden, Nat. Bur. Stand. SP-496, (1977), 423.

12. J.S. Kirkaldy, Can. J. Phys., 36 (1958) 899, 907, 917; 37 (1959) 30.

13. J.S. Kirkaldy and L.C. Brown, Can. Met. Quart., 2 (1963) 89.

14. F.J. J. VanLoo, J.A. Van Beek, G.F. Bastin, and R. Metselaar, _Diffusion in Solids, Recent Developments_ (Warrendale, PA: The Metall. Soc., 1985) 231.

15. R.F. Decker and C.T. Sims, _The Superalloys_ (New York, NY: John Wiley, 1972) 33-78.

16. S.M. Merchant and M.R. Notis, Mater. Sci. Engr. 66 (1984) 47.

17. A. Taylor and R.W. Floyd, J. Inst. Met., 81 (1952-53) 451.

18. N. Oforka (Ph.D. thesis, Sheffield Univ., 1982).

19. W.G. Moffatt, _Handbook of Binary Phase Diagrams_ (Schenectady, NY: Genium Publ. Corp, 1984).

20. M. Hansen and K. Anderko, _Constitution of Binary alloys_ (New York, NY: McGraw Hill, 1958) 1266-1267.

21. L.A. Willey, _Metals Handbook-8_ (Metals Prak, OH: Am. Soc. Metals, 1973) 260.

22. R.P. Elliott, _Constitution of Binary Alloys, First Supplement,_ (New York, NY: McGraw Hill, 1965) 254-256.

23. C.T. Sims, _The Superalloys_ (New York, NY: John Wiley, 1972) 577-588.

24. L.A. Willey, _Metals Handbook -8_ (Metals Park, OH: Am. Soc. Metals, 1973) 382.

25. G.R. Speich, _Metals Handbook-8_ (Metals Park, OH: Am. Soc. Metals, 1973) 424-426.

OXIDATION PERFORMANCE OF LASER CLAD Ni-Cr-Al-Hf ALLOY ON INCONEL 718

J. Singh, K. Nagarathnam, and J. Mazumder

Department of Mechanical and Industrial Engineering
University of Illinois at Urbana-Champaign
1206 West Green Street
Urbana, IL   61801

Abstract

The oxidation resistant materials for service at elevated temperatures must satisfy two requirements: diffusion through oxide scale must occur at the lowest possible rate and oxide scale must resist spallation. The formation of an Al_2O_3 protective scale fulfills the former requirements but its adherence is poor.   Rare earths such as Y or Hf is added to improve adhesion.   In this paper, insitu Ni-Cr-Al-Hf alloy has been developed by laser surface cladding with mixed powder feed.   A 10 kw CO_2 laser was used for laser cladding.   Optical, SEM and STEM microanalysis techniques were employed to characterize the different phases produced during laser cladding process.   Microstructural studies showed a high degree of grain refinement, increased solid solubility of Hf in matrix and Hf rich precipitates.   The paper will deal with the microstructural development during laser cladding process and its effect on oxidation.

Introduction

Alloy selection for gas turbine materials application is to ensure superior mechanical properties and good corrosion and oxidation properties at elevated temperatures. In recent years, attention has been focused on developing suitable coatings to enhance service lives and are applied over less-oxidation resistant but mechanically stronger alloy substrates. MCrAl (M = Ni, Fe, Co) systems are widely used for such coatings. These coatings tend to form Al_2O_3 rich scale which acts as a protective coating because:

1) diffusion of oxygen through the oxide scale is very slow,

2) volatility is limited, and

3) it is relatively inert in a high temperature oxidation environment.

It has been well established that the addition of trace level of reactive elements such as yttrium, hafnium, etc. to coating alloy composition greatly enhances oxide scale adherence (1-3).

Using conventional techniques, it is very difficult to incorporate rare earth (RE) elements in finely dispersed form above 1 wt%, which is the solid solubility limit (4). The inherent rapid heating and cooling rates in laser processing can be used effectively to produce an alloy with extended solid solution and uniform distribution of phases (5-8). The laser cladding technique was employed to produce Ni-Cr-Al-Hf alloy with high Hf contents compared with alloys produced by conventional techniques. The over all composition and microstructure of the laser clad materials was determined by the degree of mixing, cooling rate, and other laser processing variables (5).

The objective of this study was to develop an alloys of Ni-Cr-Al-Hf with increased Hf content which can provide better high temperature oxidation resistance properties. To the authors best knowledge, no one has reported the development of such coatings by laser cladding process.

Experimental Procedure

Inconel 718 plate was used as a substrate material. Ni, Cr, Al, and Hf powders in the ratio of 10:5:1:1 by weight were used as mixed powder feed for cladding. This gives a nominal composition of the powder mixture as 58 wt% Ni, 29 wt% Cr, 6 wt% Al and 6 wt% Hf. The cladding treatments were carried out using an AVCO HPL 10 kW CW CO_2 laser with F7 cassegrain optics (as shown in Fig. 1). The laser was operated in the TEM^*_{01} mode*. The beam focused by cassegrain optics was reflected downward toward the substrate by a flat mirror. The laser was operated typically at approximately 5, 6, and 7 kW. Specimens (approximately 6.0 mm thick) were traversed relative to the laser beam at a speed of approximately 6.35 and 10.6 mm per second. The powder was delivered to the area of interaction by a pneumatic powder delivery system with a feed screw. The powder flow was regulated by varying the speed and changing the size of the feed screw. The flow rate for the present study was approximately 0.2 gm/second. Argon gas with a flow rate of 0.017 lb./s was used to maintain a steady powder flow through the copper tubing leading to the substrate. Powder was placed on the substrate just prior to its contact with the laser.

---

*Donut-shaped laser beam with a Gaussian power distribution in the outer ring and none at the hole of the donut.

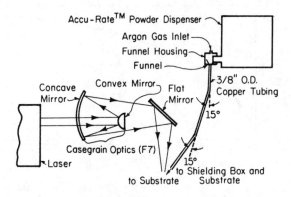

Figure 1 - Optics used for laser cladding.

An argon gas shield within a shielding box was used to minimize surface contamination during laser processing. The argon gas which directed the powder flow into the molten pool of substrate was also used in plasma suppression at the laser-substrate interaction point. The function of the shielding box was to provide the inert environment near the laser-substrate interaction point which also reduced clad porosity.

Microstructural observation of clad specimens was carried out by optical and transmission electron microscopy. After mechanical and chemical polishing to a thickness of 0.3 mm, 3 mm discs were punched from the clad material. Specimens for the TEM observations were prepared by the jet polishing technique with an electrolyte of 200 c.c. 2-Butoxy-ethanol, 400 c.c. methanol, 50 c.c. perchloric acid at 35 volts with -15°C temperature of electrolyte. Samples were observed using a Phillips 430 microscope (attached with an EDAX) operated at 300 kV.

Oxidation tests of clad material and In 718 were performed on a Thermo Gravimetric Analyzer (TGA) of Dupont 1090 system. Test was carried out up to temperature 1050°C with air flow rate between 45-50 cc/min for a duration of 60 minutes. The objective of the test was to obtain relative oxidation properties of the clad materials and In 718. In testing clad material In 718 substrate was ground off.

Results

Figure 2 represents a plot of weight gain with function of time for a period of 60 minutes at three temperatures (700, 900, and 1050°C) of the laser clad samples (laser power 5 kW and traverse speed 6.35 mm/sec). It revealed that there is parabolic weight gain of the sample at oxidation temperature 700°C, and then it decreases at oxidation temperature 900°C. There was no significant change in the weight of sample observed at high oxidation temperature 1050°C. Figure 3 represents the oxidation data of the laser clad sample operated at different laser power and traverse speed. It shows that the maximum weight gain was observed in the laser clad sample with 5 kW laser power and 6.35 mm/sec traverse speed (with over focussed laser beam) compared with the other samples of 5, 6, and 7 kW laser power,

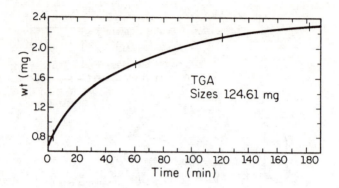

Figure 2 - TGA plot of laser clad sample
at three temperatures 700°C, 900°C, and
1050°C with function of time.

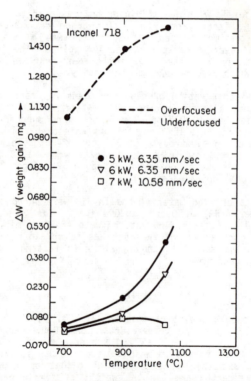

Figure 3 - A plot of weight gain
versus temperature for different
laser processing conditions.
All samples were of the same size.

and traverse speed of 6.35 and 10.58 mm/sec (with under focused laser beam). Figure 3 also revealed that keeping the traverse speed constant (i.e., 6.35 mm/sec), with increasing laser power for under focused laser beam results in a decrease in weight gain. Table I gives the oxidation data for Inconel 718 at oxidation temperature 900°C for 1 hour. On comparing the oxidation data, the laser clad alloys show better oxidation resistance properties. In addition, the alloys prepared by under focused laser beam shows a dramatic increase in the oxidation resistance properties as compared with the substrate and alloys prepared by over focused laser beam. This will be discussed more in the next section.

The electron probe microanalysis revealed that the average composition of each constituent element is more or less uniform throughout the clad region except Hf. The fluctuation of Hf is due to the presence of three different Hf beaming phases: undissolved Hf, Hf rich precipitate and solid solution of Hf in the matrix. Depending on the location of the microprobe relative to any of the above phase a large fluctuation was observed. Composition of the phase with extended solid solution of Hf was found to be uniform throughout but two other phases exhibited large fluctuation with respect to the composition of the matrix and in these phases other alloying elements also change accordingly. This will be discussed in more detail in the next section.

On comparing the oxidation properties of the substrate with laser clad samples, it was found that all laser clad samples showed better oxidation properties and especially under focused laser clad samples at oxidation temperature 1050°C. In order to understand the high temperature oxidation behavior of these alloys, microstructural investigation of the laser clad samples were carried out. The optical micrographs of the laser clad sample are presented in Fig. 4. It shows uniform distribution and formation of second phase in the laser clad matrix. As there was not any sharp interface boundary between the laser clad region and substrate matrix, it suggests a good mixing and strong metallurgical bonding between the laser clad region and substrate. Figures 5a and b are the optical and SEM micrographs of the laser clad sample (at 5 kW, 6.35 mm/sec with over focus laser beam) reveals the uniform distribution of undissolved Hf and Hf rich precipitates throughout the matrix. The segregation of undissolved Hf particles was not observed at the grain boundaries. By changing the laser processing conditions (i.e. laser power 6 kW and traverse speed 6.5 mm/sec with under focused laser beam), the size and volume of second phase precipitate were also changed, Fig. 5c and d. Relatively large volume fraction with a fine distribution of second-phase precipitate as well as undissolved Hf particles were observed. This is due to mainly laser optics interaction with the substrate during melting which would be discussed in the next section. To see the detailed internal structure of the laser clad region, STEM investigation was carried out. A general survey of the thin foils show fine and uniform distribution of undissolved Hf particle as marked by arrows, and also Hf rich precipitates in the nickel matrix (Fig. 6a and b). In addition, very fine gamma prime (γ') precipitates were also observed in the matrix (Fig. 6c). The average size of γ' precipitate is about ~70 A°. The γ' precipitates are coherent with the matrix and can only be seen in the dark field micrograph using one of the γ' precipitate super lattice reflections (Fig. 6c). The average composition of the matrix is displayed in Fig. 7. Figure 8 displays an electron micrograph of the laser clad region showing the precipitates containing high density of stacking faults, (Fig. 8a and b) and corresponding diffraction pattern from the precipitate is presented in Fig. 8c. The average composition of precipitate is presented in Fig. 9, which reveals that the precipitate is rich in Hf. As the equilibrium phase diagram of the Ni-Cr-Al-Hf quaternary system is not available and often metastable phases are formed during rapid solidification process (5), so it

105

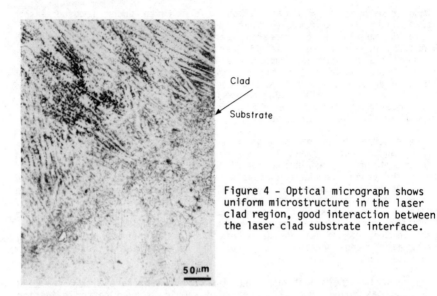

Clad

Substrate

Figure 4 - Optical micrograph shows uniform microstructure in the laser clad region, good interaction between the laser clad substrate interface.

50μm

is difficult to comment about the exact crystal structure of this Hf rich phase. As the precipitates contains mainly Hf and Ni elements with very small amount of Cr and Al. thus pseudo binary Ni-Hf diagram can be considered. Based on the diffraction analysis (Fig. 8c) and microchemistry of the precipitate (Fig. 9), it appeared that the precipitate is an hexagonal crystal structure (5).

Discussion

It has been observed that inherent rapid melting and solidification during laser cladding process produces very fine grained microstructure along with increased solid solubility of Hf in matrix. The clad alloys were found to have better oxidation resistance properties as compared with the substrate. This is due to the combined effect of increased solid solubility of Hf, Hf rich precipitate and undissolved Hf in the Ni matrix (Fig. 3 and Table I). At this stage, it is difficult to comment about which particular mechanism or groups of mechanisms are responsible for improving high temperature oxidation resistance properties, as various mechanisms were proposed by various authors for the MCrAlRE alloys (9-14). The most beneficial effect of the laser cladding process was fine microstructure, uniform distribution of Hf rich precipitate and undissolved Hf particles. In addition, the undissolved Hf was in the form of hafnium and not the oxide of hafnium i.e., hafnia (5). This was confirmed by the selected area diffraction pattern obtained during transmission electron microscopy. During oxidation, the undissolved Hf will form its oxide which will probably act as a pegs, which will increase the bonding between the oxide scale and substrate (15). Similarly, Hf rich intermetallic precipitates would decompose at high temperature and would form its oxide during oxidation process. Therefore, there would be always a competitive reaction for the formation of oxide of hafnium among the hafnium present at different states. It is believed that these reactive-element oxide colonies provide excellent sinks for excess vacancies in the metal lattice that may have otherwise condensed at the scale/metal interface (16). Condensation of vacancies at the scale/metal interface would promote spallation of the

Table I   Oxidation Data for Inconel 718
(60 minutes at 900°C)

No. of OBS	Initial Wt., grams	Final Wt., grams	Wt. Change, grams
1	2.2856	2.2843	-1.3×10^{-3}
2	0.6618	0.6594	-2.4×10^{-3}

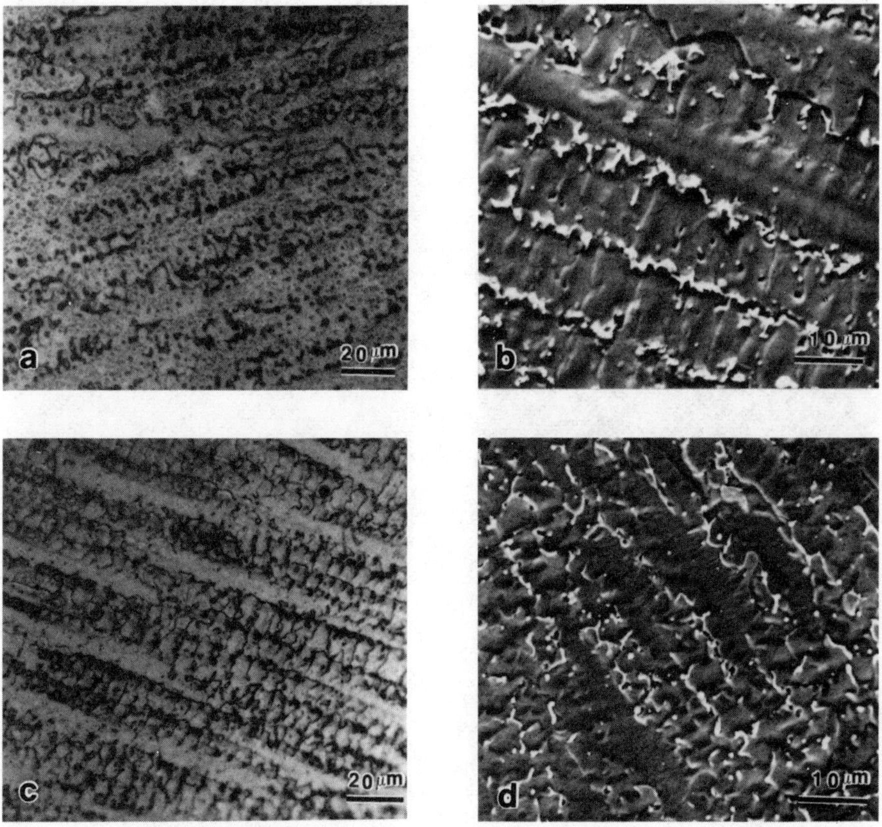

Figure 5 - Micrographs of laser clad regions for over-focused
(a) and (b) and under-focused (b) and (c) laser beam
(a)   Optical: showing uniform distribution
      of second phase
(b)   SEM: showing distribution of Hf rich
      precipitates and undissolved Hf particles
(c)   Optical showing relatively large volume
      fraction of second phase
(d)   SEM: showing fine and large volume fraction of
      second phase as well as undissolved Hf particles

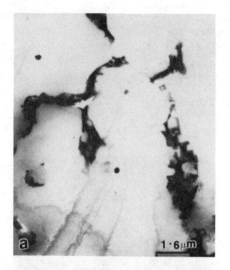 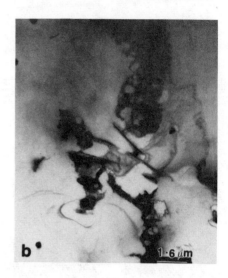

Figure 6 - General survey of TEM sample
of over focused laser clad alloy showing
(a) and (b) undissolved Hf particles
and Hf rich precipitates (c) dark field
showing that matrix is having
Ni_3Al (γ') type precipitate.

scale. Addition of Hf to the Ni based superalloys, not only improves the
oxidation resistance properties, but it also improves the mechanical properties (especially creep properties) and controls the growth of gamma prime
precipitates during high temperature cyclic treatments (17).

Theoretical calculations have shown that, in nickel based alloys containing aluminum with RE additions (Y or Hf), the RE atoms have a strong
binding energy to the aluminum atoms, because of charge transfer from their
high-lying valence orbitals to lower-lying Ni s-d bands (18). Thus, the
outward diffusion of cations would apparently be suppressed because Cr atoms
are tied up by RE (Hf) atoms. Therefore, the mechanism of scale growth may
be controlled by the inward transport of oxygen. Further, as mentioned
above, that the reactive element provides an excellent sites to condense
excess vacancies, it also acts as a nucleation site for the continuous
formation of Cr_2O_3 layer thus prevent the depletion of Cr content in the
alloy adjacent to the interface (18).

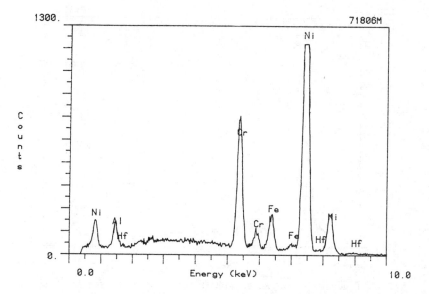

Figure 7 - STEM x-ray microanalysis from the matrix.

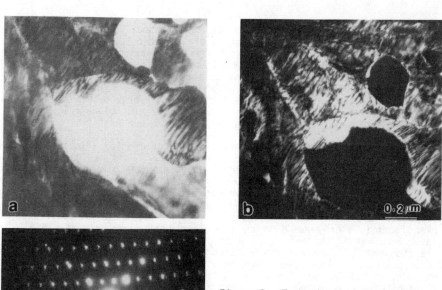

Figure 8 - Transmission electron micro graph of the laser clad sample showing the meta stable phases (a) bright field (b) dark field (c) corresponding diffraction pattern.

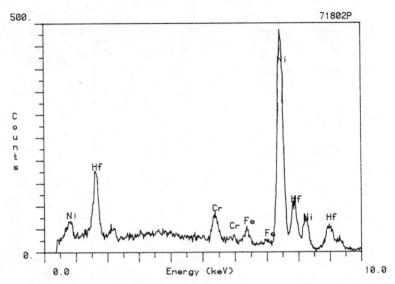

Figure 9 - STEM x-ray microanalysis from the second phase showing that the second phase is rich in Hf with small amount of Ni, Cr, and Al.

At this stage, it is difficult to comment about the extent of contribution of the size and volume fraction of γ' precipitate present in matrix towards the oxidation resistance properties. It has been observed that the volume fraction and size of γ' precipitate increases with increase in laser power or decrease in the traverse speed (19). In the γ' (Ni_3Al) precipitate, Al would be replaced by Hf where as Ni would be replaced by Cr, Co, Fe, etc. (5). In order to have good binding energy between Hf with Al, Ni, Cr, it would be an ideal to have optimum size and volume fraction of γ' precipitate present in the matrix.

The oxidation properties of under focused laser clad alloys were found to be better than over focused laser clad alloys. This was due to concentration difference of alloying elements present in a given volume of melt pool region which would be discussed in detail as follows. The laser beam diameter (3 mm) was the same as over focused and under focused position (Fig 10a), but their interaction with the substrate was different. Figure 10a represents a simple laser beam optics. When the substrate was positioned at over focused laser beam location with 3 mm beam diameter (Fig. 10b), the laser beam was in the converging mode. Once the laser beam was incident on the substrate, it would like to penetrate inside the substrate up to the focal point. The laser beam would melt the substrate up to focal point i.e. depth d_1 of substrate and mix with melted powder to form a clad layer. If the substrate was located at the under focused laser beam with 3 mm diameter (Fig. 10c), the interaction of laser beam with substrate would be always with diverging beam with lower power density. The depth of penetration of laser beam inside the substrate would be low (d_2) and would mix with same volume of melted powder to form a clad layer. Since the melt depth (d_1) for over focused laser beam was more than d_2 for under focused condition, melted volume for the over focused laser beam will be higher. This suggests that laser clad alloy produced by over focused laser beam would be more diluted with alloying elements such as Cr, Al and Hf. This will also directly control the total volume fraction of Hf rich precipitates formed, solid solu-

110

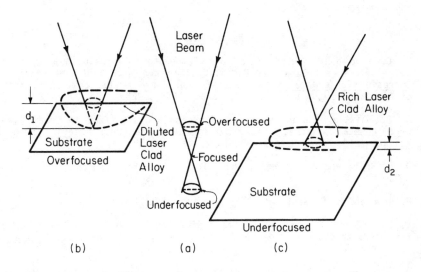

Figure 10 - Laser optics occurring during laser cladding process.

bility of Hf in the matrix and undissolved Hf particles present in the total volume of solidified melt pool region. Figure 5 supports the above statement that total volume fraction of Hf rich precipitates and undissolved Hf particles are more in case of under focused laser clad alloys as compared with over focused laser clad alloys because they were rich in alloying contents present in the less solidified melt pool region. Therefore, better oxidation properties are expected for laser clad alloys produced by underfocused laser beam.

Summary

By the laser cladding process, a fine microstructure with uniform distribution of undissolved hafnium, Hf rich intermetallic precipitates and a matrix with extended solid solubility of Hf in the laser clad region was achieved. The oxidation resistance properties of the laser clad alloy produced by under-focused laser beam was found to be better than the alloy produced by over-focused laser beam as well as Inconel 718 substrate. It is postulated from the present observation that the reactive elements would change the mechanism of the scale formation as a competitive reaction is taking place for the oxidation. These reactive oxidized particles will act as sinks for excess vacancies thus inhibiting their condensation at the scale/matrix interface and possibly enhancing scale adhesion.

Acknowledgment

This work was made possible by a grant from the Air Force Office of Scientific Research (Grant No. AFOSR-85-0333). The authors wish to thank Dr. Rosentein and Major Hagar of AFOSR for their continued encouragement. The studies involving electron microscopy were performed in the Center for Microanalysis of Materials in the MRL of University of Illinois at Urbana-Champaign.

References

1. G. C. Wood and F. H. Stott, "High Temperature Corrosion," Rapp R. A. ed., Conference held at San Diego, California, March 1981, International Corrosion Conference Series, NACE-6, published by National Association of Corrosion Engineers, Houston, Texas (1983), 227.

2. D. P. Whittle, and D. H. Boone, "Surface and Interface in Ceramic and Ceramic-Metal Systems," Pask, J. and Evans, A. ed., Material Science Research, 14, Plenum press, New York, 487.

3. D. P. White and J. Stringer, Phil. Trans. Roy. Soc. London., A295 (1980) 309.

4. P. Nash and R. F. West, Metal Science, 15 (1981) 347.

5. J. Singh, and J. Mazumder, "Effect of Extended Solid Solution of Hf on the Microstructure of the Laser Clad Ni-Fe-Cr-Al-Hf Alloys," Acta Met. (submitted).

6. J. Singh and J. Mazumder, " Microstructural Evolution in Laser Clad Fe-Cr-Mn-C Alloys," Material Science and Technology, 2 (1986) 709.

7. T. Chande and J. Mazumder, J. App. Physics, 57 (1985) 2226.

8. I. M. Allam, D. P. Whittle, and J. Stringer, "Corrosion and Erosion of Metals," K. Nateson (ed.) 103-117, published by TMS-AIME, 1980.

9. J. G. Smeggil and A. W. Funkenbusch, "A Study of Adherent Oxide Scale," United Technologies Research Center Report #R85-916564-1, May 1985.

10. D. P. Whittle and J. Stringer, Photos. Trans. R. Soc. Lond., Ser. A295, (1980) 309.

11. J. K. Tien and F. S. Pettit, Metal Trans., 3, (1972) 1587-1599.

12. F. A. Golightly, F. H. Stott, and G. C. Wood, Oxide Met., 10, (1976) 163-187.

13. J. E. Antill and K. A. Peakall, J. Iron Steel Inst., 205, (1967) 1136-1142.

14. H. Pfeiffer, Werkst. Korros., 8, (1957) 574.

15. W. E. King, N. L. Peterson, and J. F. Reddy, J. of Physics (1985) 423.

16. J. G. Smeggil, A. W. Funkenbusch, and N. S. Bornstein, High Temperature Science, 12 (1985) 163.

17. E. L. Hall and S. C., Huang, Met. Trans., 17 (1986) 407.

18. H. Yang, G. E. Welsch, and T. E. Mitchell, Mat. Sci. and Eng. 69, (1985) 351.

19. J. Singh and J. Mazumder, unpublished research work.

ELEVATED TEMPERATURE EROSION EVALUATION OF

SELECTED COATINGS PROCESSES FOR STEAM TURBINE BLADES

J. Qureshi

Westinghouse Electric Corporation
Orlando, Florida 32826-2399

A. Levy and B. Wang
University of California, Berkeley, California 97420

Abstract
The exfoliated scales from boiler turbine causes hard particles that erode steam turbine components, particularly stationary and rotating blades. To select an appropriate erosion resistant protective coating for fossil steam turbine blades, six promising coating processes comprising ten different coatings were selected for evaluation by erosion testing, metallography, and mechanical tests of coated blade materials. This paper describes the results of the elevated temperature erosion tests and metallographical studies and discusses how erosion resistance is related to coating morphology.

I. Introduction

The principal erodent source for the erosion of high-pressure turbine blades and valve components in fossil-fired power plants is hard particles which are magnetite (Fe_3O_4) scales formed at elevated temperature by reaction of steam with steam generator materials. During boiler transients (startup and cooldown cycle), these magnetite scales exfoliate from the interior surfaces of boiler tubes and are carried as hard particles in the steam flow to the turbine. This solid particle erosion (SPE) attacks the blade airfoil at various impingement angles, resulting in damage at the airfoil trailing and leading edge and concave side of stationary/rotating blades. This damage is more pronounced in the impulse stage blade than the reaction blade because the steam velocity is high and solid particles are centrifuged against the impulse blade airfoil at low impingement angles (20°-30°).

The turbine blade damage will reduce turbine efficiency and also increase turbine maintenance cost; consequently, the turbine operational cost will increase. The results of an EPRI study indicate that the average yearly cost of SPE damage is $0.070/kW.[1] Moreover, an ASME-ASTM-MPC study found total SPE damage in the U.S. to cost 100 million dollars per year.[2] In order to solve this erosion problem, the turbine manufacturers and utilities are trying various methods, such as preventing the scale formation in the boiler,[3] trapping the particles before they enter the steam turbine,[4] and armoring the blades by use of a protective layer.[5-7] Each method has its own merits and faults. Since significant metallurgical advances in erosion and wear protection through the use of special surface treatments have been successfully demonstrated in the aircraft engine,[8] petrochemical, and geothermal industries, Westinghouse is evaluating the surface treatment as a method to solve this erosion problem.

Experienced commercial suppliers of candidate coating systems were selected and coated test specimens were procured. Elevated temperature erosion tests were conducted to determine the behavior of the protective coatings on type AISI 403 and AISI 422 martensitic stainless steels. These tests ranked the erosion resistance of the various coatings systems (materials and processes) that were evaluated and determined a relationship between the composition, morphology, and defects in the coating and their erosion rates. To select the appropriate erosion resistant coating for Westinghouse steam turbine blades, the evaluation was based on erosion resistance, microstructure, fatigue, and creep strength of coated alloys and the coating quality on actual blades. This paper focuses the erosion and metallographic analysis results obtained on coated and eroded specimens.

II. Selection of Coating Process and Material

To select the appropriate coating processes and materials for this turbine blade program, a computerized literature search for information on erosion resistant coatings was conducted. The search indicated that there were several specialized coating materials and processes that have been applied on different substrates and successfully used in various industrial applications.

114

Aluminide and boride diffusion coatings have successfully reduced the corrosion and erosion of compressor blades. Boiler manufacturers are applying a chromized diffusion coating to reduce scale formation in boiler tubing.[3] Westinghouse diffusion coated high-pressure (H.P.) stationary nozzle vanes in 1980 which were exposed to service conditions for three years. The three year service experience indicated that the diffusion coating had doubled the vane life.[6]. Since then, many H.P. nozzle vanes have been boride diffusion coated to increase the erosion resistance of the stationary vanes. Therefore, this diffusion coating was a good candidate for this rotating blade program. Spray coatings have also been tried on Steam Turbine blades to reduce erosion.

General Electric[9] and Southern California Edison[7] plasma sprayed steam-turbine blades to reduce solid particle erosion. Since 1980, General Electric has plasma sprayed tungsten carbide on 38 rows of blades for 17 utilities[9]. The performance of tungsten carbide coating is variable[9]. Moreover, General Electric evaluated various spray coating materials and concluded from lab data that chromium carbide in the heat-treated condition had the best erosion resistance among the coatings tested. Based on this experience, the spray coatings process and the appropriate materials (WC and Cr_3C_2) were included in this blade coating program.

Based on the available field data and laboratory experience in various applications and, in particular, the ability of the process to coat steam turbine blades without affecting blade mechanical properties, six processes consisting of ten coating materials and ten coating vendors were selected. Table 1 shows coating materials, processes, and characterization data for all the coating systems evaluated.

III. Test Conditions And Equipment

Since the blade material is either AISI 403 or 422 stainless steel, flat test coupons of both steels were prepared and coated with the above selected materials and processes. All the coated test coupons were prepared to a surface finish of 0.1575 μm (63 μ in) rms and penetrant inspected before the erosion tests. To evaluate the effect of particle material size, velocity, impingement angle, and gas environments, these coatings were erosion tested at two conditions. Some coatings were deleted or improved as the erosion test progressed from Condition A to Condition B.

		Condition A	Condition B
Test temperature	(°C)	538° (1000 F)	382° (720°F)
Environment		Air	Steam
Particles		Chromite	Magnetite
	powder	powder	
Particle velocity	(m/s)	152	229, 305
Particle size	(μm)	75	40
Impingement angle	(deg)	30°, 90°	22.5°, 90°
Test time	(min)	30	90
Total mass of particles impacting specimen(g)		25	750

115

IV. Results

1. Erosion Rates

The results of condition A erosion tests on the candidate coatings on AISI 403/422 stainless steels (SS) are listed in Table II and Figure 1. Coating No. 9 was not deposited on type 403 SS, and hence test data do not appear in Table II and Figure 1 for that configuration.

Table I. Candidate Coatings Applied on AISI 403 and AISI 422SS

Test Material Identification	Coating Process	Coating Material	Surface Hardness HV-500g	Coating Thickness (μm)	Material Density g/cm³
0(403)	None	None	225	0	7.75
0(422)	None	None	275	0	7.75
1	Diffusion	Nitrided case	1027	254	7.1
2*	Diffusion	CrB + FeB	1283	51	7.13
3	Detonation gun®ᵈ	Cr_3C_2	1009	114	6.4
4	Physical vapor dep.	TiN	713	15	5.2
5	Plasma Spray	NiCrBC	927	127	12.3
6	High-velocity plasma spray	WC + Co	1006	114	13.8
7	Electroless nickel	Ni + P	617	102	7.85
8	Conforma Clad overlay®ᵈ	WC + NiCrB	982	127	11.65
9	Chem. vapor deposition	WC	713	114	19.35
10	Electrospark dep.®	Cr_3C_2	613	25	7.12
11	Electrospark dep.®ᵈ	Co-Mo-Cr	615	25	8.64
12*	Diffusion	CrB + FeB	...	71	7.13

ᵃPatented processes of coating vendors.

* Coatings No. 2 and No. 12 are same materials but were applied by different vendors.

Table II. Erosion Ranking of Tested Coatings on 403/422SS

Coating Process	Test Material Test No.	Ranking On 403 Stainless Steel At Impingement Angle 30°	Ranking On 403 Stainless Steel At Impingement Angle 90°	Ranking on 422SS At Impingement Angle 30°	Ranking on 422SS At Impingement Angle 90°
None	0	9	—	10	7
D gun	3	1	3	1	3
Conforma Clad®	8	2	4	2	4
Boride diffusion	12	3	1	3	1
CVD	9	b	b	4	5
PVD	4	4	5	5	6
Boride diffusion	2	5	2	6	2
High-velocity plasma spray	6	6	6	7	9
Plasma spray	5	7	8	8	8
Gas net diffusion	1	8	7	9	10
Electrospark deposit	10	a	a	a	a
Electrospark deposit	11	a	a	a	a
Electroless nickel	7	a	a	a	a

[a]Coating wore through.
[b]Not tested.

The results of tests at condition A are reported herein. The results of tests at condition B are not yet available. An elevated temperature erosion tester at the Lawrence Berkeley Laboratory (LBL) described in Ref. 12 was used. Erosion rates were determined by periodic weight loss measurements using an analytical balance that measured specimen weight to an accuracy of ±0.001g. After testing, the specimens were observed in a scanning electron microscope having an energy dispersive x-ray analysis device. Steady-state erosion rates were achieved for all the tested materials and were ranked.

The volume loss was calculated by dividing the measured weight by the density listed in Table I and shown in Figure 1. The materials are listed in the order of their performance at impingement angle $\alpha = 30°$ except that bare 422SS/403SS is listed first so the improvements in erosion behavior due to the various coatings can be readily compared.

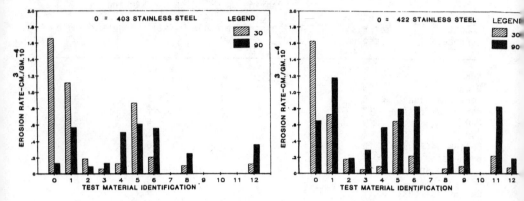

Figure 1   Elevated temperature erosion rate of various coatings on AISI
403/422SS (coating Nos. 7, 10. and 11 wore through).

The rankings differed between the α = 30° and 90° tests.   Also, since
three specimen (Nos. 7, 10, and 11) had the coatings completely removed in
the α = 30° tests, these materials cannot be ranked.   The ranking at 30°
on 403 and 422 stainless steel is the same, and ranking at 90° was
comparable for both steels.

2.Metallography

The morphology of the coatings in the eroded area generally correlated
quite well with the erosion rates measured.   In general, the rankings of
the coatings on each of the two steel substrates were comparable.
Therefore, the metallographic analysis concentrated on coatings on the
422SS.   Only selected micrographs for each coating were included in this
paper.

3.Bare Stainless Steel

The surfaces of 422SS specimens eroded in the laboratory tester at impact
angle α = 30° and 90° are shown in Figures 2 and 3, respectively.
Microexamination indicated a rippled surface which is typical for ductile
metals eroded at shallow impact angles.   The higher magnification photo of
the surface shows a fine distribution of platelets, shallow craters, and
narrow gouges.   The size of microstructural elements can be related to the
fine microstructures of the tempered martensite of 422SS.   There is no
evidence of oxide scale on the surface.   The energy dispersive x-ray EDX
peak analysis is typical of a 400 series stainless steel.

 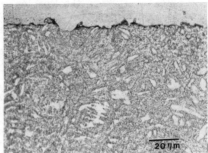

Figure 2   Surface and cross section of 422SS eroded at 30°

The surface of the 403SS specimen at $\alpha = 90°$ indicates a considerable embedment of the erodent particles.   The energy dispersive spectra EDS peak analysis indicates that constitutents of the erodent are now contained in the metal surface.   The thin, uneven layer of erodent was found in the cross section.   The lower erosion rate of 403SS at $\alpha = 90°$ (shown in Figure 1) could be due to some level of protection that was provided by the embedded particles on the ductile metal.

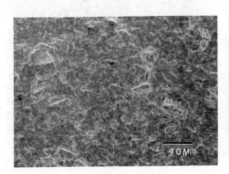 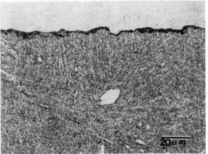

Figure 3   Surface and cross section of 422SS eroded at 90°

4. Spray Processes

The (D gun) coating (No. 3), which was the best performing coating at $\alpha = 30^{\circ}$, is shown in Figure 4. The uneroded, sprayed material consists of very fine grain particles with very little porosity between them (Figure 4). When the coating was eroded at $\alpha = 90^{\circ}$, it was further smoothed and some of the voids were filled with eroding debris. Some of the small voids in the sprayed coating had been opened up and the smooth, ground surfaces had been roughened. At $\alpha = 90^{\circ}$, the eroded surface at the center was rougher than that which was observed in the $\alpha = 30^{\circ}$ test. The fine grain structure of the coating was found in both the center and outer zone micrographs. Both had basically the same morphology and were very similar to the uneroded coating's surface. In the erosion of brittle materials, the finer the grain size and the lower the level of porosity, the lower is the erosion rate because only small pieces can be chipped out by particle impacts.[13] The excellent performance of this coating is directly related to the fine-grains, low porosity morphology of the coating. It is eroded by cracking and chipping off small pieces whose size was determined by the basic grain size of the coating.

The high velocity plasma spray coating (No. 6) is typical of a good quality plasma spray coating. The reason that it did not rate higher in its erosion behavior is the fine network of pores that can be seen in the cross-section micrograph in Figure 5 and in the surface micrograph of the outer zone of the specimen edge. In earlier work on the erosion of plasma spray coatings at LBL,[14,15] it was determined that the even distribution of almost equiaxed pores throughout plasma spray coatings was more detrimental to their erosion resistance than the more planar pores that occur between layers of a flame spray coating. The evenly distributed, equiaxed pores expose unsupported matrix material that can easily be broken off by impacting erodent particles.

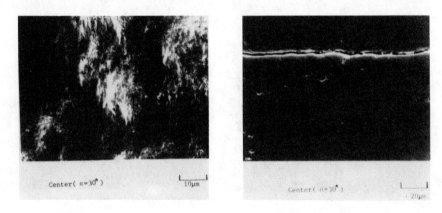

Figure 4. D gun spray coatings (No. 3) shows fine pores and eroded surface at 30°

The morphology of the eroded surfaces at both $\alpha = 30^0$ and 90° are similar. In the center zone, the debris tended to smear over and fill in the voids. The coatings were removed from the central area, but the outlying areas where fewer erodent particles impacted still show evenly distributed pores in the coating. The EDX analysis indicates that the coating was essentially WC. The size as well as the geometry of these pores is directly related to the erosion rates of the sprayed coatings. The detonation gun coating (see Figure 4) had finer as well as fewer pores and, therefore, eroded at lower rates than plasma spray coatings. For coatings that do not have crack networks, the less porous the coating is and the smaller the pore sizes are, the greater is their erosion resistance.[14]

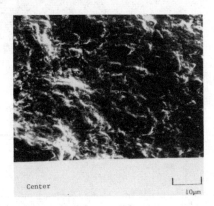 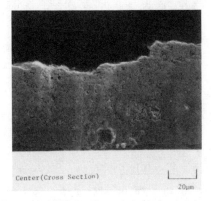

Center

10μm

Center(Cross Section)

20μm

Figure 5 High-velocity plasma spray coating (No. 6) shows pores and eroded surface at 30°

The plasma spray coating identified as No. 5 in Table I had both quite large as well as small pores and a crack network. It differed from plasma spray coating No. 6 which only had small size pores. The mechanism of material loss as the result of the large pores and cracks is shown in Figure 6. The large chunks of coating material (hard particle) chipped off from the surface, which resulted in high erosion rates. The difference in pore size between coatings 5 and 6 had a great effect on their erosion rates. This can be seen by comparing erosion rates of the two plasma spray coatings in Figure 1. At $\alpha = 30^0$, the unsupported sides of the larger pores of coating No. 5 can be more readily knocked off the surface by eroding particles than they can at $\alpha = 90^0$. Therefore, Table II shows a difference in ranking between the two plasma spray coatings at $\alpha = 30^0$, while essentially minimum difference in rates occurred in the $\alpha = 90^0$ tests. Generally, the morphology of eroded materials can be closely related to their erosion rates.

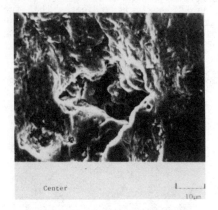

Figure 6. Plasma spray coating (No. 5) shows large pore and eroded surface at 30°

5. Flexible Overlay

The surface of the eroded Conforma Clad coating in Figure 7 shows a fine grain that was ground after cladding. The cross section and the EDX peak analysis indicates that the material consisted of grains of WC of various sizes in a matrix of NiCrB alloy. A few voids of reasonable size were present as were some cracks. The surface and cross-section micrographs of the center zone from the α =90° test shows that the matrix was preferentially eroded, leaving WC particles projecting at the surface.

The α =30° test specimen surface and cross section shown in Figure 7 indicate that both the WC particles and the NiCrB matrix alloy eroded at nearly the same rate. The WC appeared to provide some protection for the matrix as the erosion rate at α =30° was much lower than at α =90° (see Table II). In this flexible overlay (coating No. 8), the dense coating structure (as also found in the highly ranked D gun coating No. 3) together with a brazed alloy results in low erosion rate. The widely dispersed voids and cracks did not result in a major reduction in the coating's performance.

122

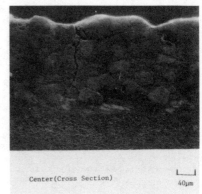

Surface(center)

10μm

Center(Cross Section)

40μm

Figure 7 Surface and cross section of eroded Conforma Clad coating (No. 8) at 30°

6. Vapor Deposition Processes

The chemical vapor deposition applied WC coating No. 9 has been tested before at LBL and the results of the current tests are similar to the previous results.[16] The coating No. 9 was tested in the heat treated condition. The microexamination shows [Figure 8(a)] the uneroded material was dense and had a very small grain size of about 100 Å . There appears to be a thin layer of material on top of the base, some of which has spalled off. This may be evidence of a second deposition during the coating processes to fill cracks left in the first deposited layer.

This type of dense and small grain size microstructure erodes by cracking and chipping off very small particles; hence, the low material loss rates as shown in Figure 1. The microexamination of the center and halo zones of the tested specimen indicate only a small difference in the appearance of the eroded surfaces at both impact angles. The mechanism of material loss and the small difference in the appearance of the eroded surface at $\alpha = 30°$ and 90° cannot account for the relatively large difference in their erosion rates. The microexamination shows a few cracks that appear to have been filled.

The surface layer consisted of small chips of coating that were about to be removed and some larger cracks that were parallel to the eroding surface [Figure 8(a)]. The high-density, very small grain size of the coating is evident.

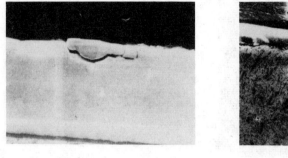

Center Cross Section(α= 30°) |___| 20μm

a

Center Cross Section(α=30°) |___| 20μm

b

Figure 8   Surface and cross section of CVD coating (No. 9) and PVD coating (No. 4) eroded at 30°

The physical vapor deposition applied TiN coating (No. 4) was only 15 μm thick.   At α = 30° its fine grain, dense morphology resisted the impacting particles well enough to leave a thin coating layer at the end of the short time test, as can be seen in Figure 8(b).   The eroded surface in the center zone (α - 30°) shows the fine grained, smooth surface that occurred when material loss was by cracking and chipping of very fine pieces.   The smoothness of the surface can also be seen in the α = 30° cross-section micrograph, which has a flake of dirt on its right-hand side.   The EDX peak indicated that the TiN completely covered the surface. The relatively low hardness of this coating indicates that it is the fine grain structure and high density that determined the erosion behavior and not the hardness.

In contrast, the surface of the α = 90° test may be indicative of the brittleness of the TiN.   If this is the case, a thicker coating may or may not have been successful.

7. Diffusion Process

The boride diffusion coatings (Nos. 2 and 12) were applied by a pack cementation process.   Figure 9 shows the coating No. 2 α = 30° specimen microstructure of the boron diffused coating which appeared to have CrB/FeB particles embedded in an Fe, Cr, B matrix layer.   The boron diffused coating had a fine grained matrix that eroded at a low rate. However, the coating had a significant crack structure extending through it that compromised the fine grained, dense matrix in its ability to resist erosion.   The EDX peaks indicate that the Cr content on the surface was greater than the average Cr content through the cross section.   This is probably due to the concentration of Cr in the CrB particles on the surface.

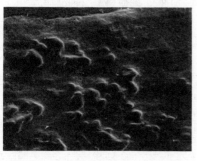
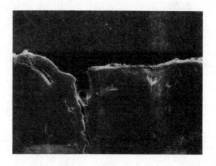

Eroded Center   Center(Cross Section)
                    20μm   20μm

Figure 9   Surface and cross section of boride diffusion coating (No. 2) eroded at 30°

This coating had the highest surface hardness 1283 HV of any of the coatings tested in the program, yet it ranked sixth in erosion resistance at α = 30° . Hardness does not appear to have a significant relationship to erosion in most of the work on erosion performed at LBL over the past ten years.[17]   A very soft material will erode more than other materials. However, if a rather low hardness threshold is achieved, it is other properties and the morphology of the material that determines its erosion behavior.   In the case of this diffused CrB/FeB coating, the crack network through the coating caused the comparatively high erosion rate.

Boride diffusion coating No. 12 is applied by the same process as coating No. 2; however, the coating parameters differ from coating No. 2.   This boron diffusion coating had a few cracks which resulted in an intermediate erosion rate.   This coating ranked better at 30° than coating No. 2 because it had fewer cracks than coating No. 2.   The fact that the coating behaved well, even with a few well-defined cracks, indicates that it may be superior performing material if the cracks can be eliminated in the deposition process.

The nitride coating No. 1 was applied by a gas nitriding process.   The nitrided case material eroded at a high rate even though it had one of the high hardnesses.   The hardness of coating material has a little or no effect on the erosion state of the coating and is shown in Figure 10.   The uneroded surface shows an evenly distributed network of pores and a crack network which contributed to its high erosion rate.   The coating eroded by a mechanism that consisted of cracking and chipping off pieces of coating.   This occurred in the same manner at both impact angles.   The outer, white layer of the nitrided case is very brittle and was completely eroded off.   The coating's morphology itself did not give the appearance of an erosion-prone coating when compared to the mophologies of other coatings tested in this series that had much lower erosion rates. It is thought that the basic reason for the high erosion rate of the nitrided 422SS is that the nitrided scale was considerably more brittle than the other coatings and could be cracked and chipped much more readily.

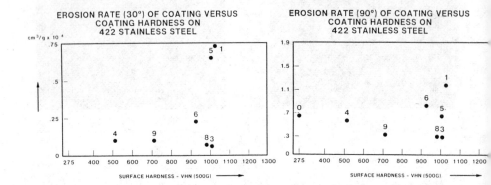

Figure 10.   Hardness Effect on the Erosion Rate of Various Coatings

8. Electrospark Deposition

The electrospark deposition (ESD) coating No. 10 is shown in Figure 11. Apparently the coating deposition process first selectively etched the base metal structure and deposited the coating material on the substrate. Since the coating was thin and had prior cracks in it, the coating chipped off from the base metal. The micrographs of the eroded center zone show that no coating was left on the surface after testing. This is a morphology of an eroding ductile metal.

The lower Cr peak in the EDX analysis of the center zone compared to the higher Cr peak in the uneroded outer zone indicates that the entire coating was removed.

Coating No. 11 is a poorer version of this sputtered coating with a much more severely cracked network. Both coatings were quite thin, as deposited.   The microexamination of eroded surface indicates that no coating remained at the end of the test

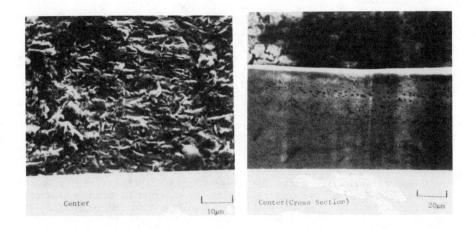

Center   10μm   Center(Cross Section)   20μm

Figure 11. The surface and cross section of ESD coating (No. 10) eroded at 30°

9. Electroless Nickel Process

The electroless nickel coating (No. 7) had large pores and a layered microstructure. It is highly susceptible to erosion in both conditions (30° and 90°). The eroded surface photograph (Figure 12) and EDX analysis indicate that no coating remained at the end of the test.

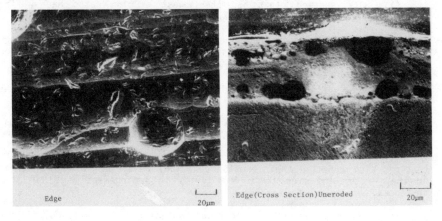

Edge   20μm   Edge(Cross Section)Uneroded   20μm

Figure 12. The eroded surface and cross section of electroless nickel coating (No. 7)

V. Discussion

The general pattern of behavior of the coatings related more to their grain size and defect structure than to their composition or hardness. The finer the grain size and the more dense the coating, the lower was the erosion rate. When the deposition process resulted in cracks in the coating, the larger and more profuse the cracks were, the higher was the erosion rate. The reason for this correlation is the basic mechanism of particle removal. The size of the chips that are formed when the impacting particles strike the surface determines how much material can be removed by a given number of particle impacts in an area of the surface. It has been observed [15] that very fine grains result in small chips and a resultant low loss rate. Larger grains and segments of coating between the pores [16] in a porous structure form larger chips when impacted, with resultant larger loss rates.

When preexisting cracks are present, such as the craze cracks that formed in several of the coatings tested, even larger pieces can be knocked off the surface by impacting particles, with resulting larger material loss rates. The three coatings which completely eroded off during the test all had extensive cracking in their as-deposited condition.

The overall hardness of the deposition of diffused coating had relatively little effect on their ability to resist erosion as shown in figure. Hardness variations shown in Table I range from 613-1283 HV_{500}. Two of the better performing coatings (Nos. 4 and 9) had hardnesses at the lower end of the range while two of the poorer performing coatings had hardnesses greater than 1000 VHN_{500}.

VI. Conclusions

(i)   The coatings that had the lowest erosion rates had a fine grained, dense microstructure with little or no cracking.

(ii)   The presence of pores or cracks such a craze cracks increased the erosion rates of coatings significantly; in the worst instance completely destroying the integrity of the coating.

(iii)   The tested coatings demonstrated higher erosion resistance at the low impingement angle (30°) than at the high impingement angle (90°).

(iv)   Four coating processes-detonation gun, Conforma Clad, chemical vapor deposition, and diffusion-had the best overall erosion resistance.

(v)   Process modifications that affected the morphology of the coatings also affected their erosion behavior, resulting in major changes in their rates and even their sensitivity to impingement angle in some instances.

(vi)   Hardness of the hard material systems had little or no relation to their erosion rates.

(vii)   The ductile materials eroded in platelet form while the brittle materials eroded by cracking and chipping of pieces.

References

[1] R. G. Brown et al., "Evaluating of ASME Solid Particle Erosion Task Group Questionnaire, " EPRI Document 2-R-83-02, 1983.

[2] R. C. Spencer "Solid Particle Erosion, What Can the Industry Afford to Spend to Eliminate SPE?", 1980 EPRI/ASME Workshop on SPE, EPRI CS-3178 Project 1885-1, 1983, pp. 2-5--2-23.

[3] I. M. Rehn "Operational Experience with Chromate Treated Superheater Tubes," EPRI Workshop on Solid Particle Erosion of Steam Turbine, Chattanooga, TN, 1985 EPRI CS4683, p. 3-15 - p. 3-27.

[4] C. Verpoort, C. Maggi, and H. Bartsch, :The Protection of Steam Turbines Against Solid Particle Erosion," EPRI Workshop on Solid Particle Erosion of Steam Turbine, Chattanooga, TN, 1985 EPRI CS4683, p. 4-19-4-41.

[5] W. J. Simmer, J. H. Vogan, and R. J. Lindinger, "Reducing Solid Particle Erosion Damage in Large Steam Turbines," American Power Conference, Chicago, IL., 1985.

[6] L. D. Kramer, J. I. Qureshi, R. A. Rousseau, and R.J. Ortolano, "Improvement of Steam Turbine Hard Particle Eroded Nozzles Using Metallurgical Coatings", ASME Paper No. 83 JPGC-PWR-29.

[7] R. L. Ortolano, "Resisting Steam Turbine Abrasion Damage by Using Surface Treatment System," ASME Power Generation Conference, 1985.

[8] J. Newhart, "Evaluation and Controlling Erosion in Aircraft Turbine Engines," publication of Haret Air Propulsion Test Center, 1983.

[9] S. T. Wlodek, "Development and Testing Plasma Spray Coatings of Solid Particle Erosion Resistance", EPRI Workshop on Solid Particle Erosion of Steam Turbine, Chattanooga, TN, 1985.

[10] E. F. Sverdrup, et al., "Control of Erosion in Power Plant Fans," Final Report No. EPRI 1649-4, 1983.

[11] R. Johnson, "Electric Spark Deposition--A. Technique for Producing Wear Resistant Coatings", Proceedings of the International Conference on Wear of Materials (Vancouver, Canada, 1985).

[12] A. Levy and Y. F. Man, Wear (to be published).

[13] A. Levy, T. Baker, E. Scholz, and M. Aghazadeh, "Erosion of Hard Metal Coatings", Proceedings of High Temperature Protective Coatings Conference (The Metallurgical Society, AIME, city 1983), p. 339.

[14] G. Hickey, D. Boone, A. Levy and J. Stiglich, Thin Solid Films 118,321 (1984)

[15] T. Foley, and A. Levy, Wear 91,45 (1983).

[16] A. Davis, D. Boone, and A. Levy, Wear (to be published).

SIMULTANEOUS CHROMIZING-ALUMINIZING OF IRON AND IRON-BASE ALLOYS BY PACK CEMENTATION

Robert A. Rapp, Dayung Wang, and Tim Weisert
Department of Metallurgical Engineering
The Ohio State University
Columbus, Ohio 43210

ABSTRACT

Iron and a low-alloy steel were simultaneously chromized and aluminized by means of a pack cementation process. A masteralloy of high Cr/Al activity ratio is essential for deposition of both Cr and Al. Therefore, a 95Cr-5Al (by weight) alloy was used as the masteralloy for coating a pure iron substrate, and a 90Cr-10Al alloy was suitable for a 2.25Cr-1Mo steel. A mixed activator salt of $1NaCl:2AlCl_3$ provided the proper ratio of volatile Cr- to Al-halides to achieve the desired surface composition. A rotation (tumbling) of the pack enhanced metal deposition by eliminating the depletion zone around the substrate. The resulting diffusion coatings had a Kanthal[*]-like surface composition, Fe-20Cr-4Al, which provides superior high-temperature oxidation resistance. Thermogravimetric oxidation testing (TGA) in air at 1000'C confirmed the excellent oxidation resistance of this coating.

Preliminary studies showed that yttrium could also be introduced into the coated surface by direct mixing of fine yttria powder into the cementation pack. Very limited testing showed a significant increase in the oxide adhesion for cyclic oxidation.

[*]Trademark, Kanthal Corporation.

I. INTRODUCTION

At high temperatures (700-1200 C), materials in utility boilers, fuel cells, petrochemical plants, etc., are subject to various corrosion problems. To obtain the best combination of high-temperature strength and oxidation resistance, strong and fabricable substrate alloys can be coated with Cr, Al or Si to produce thin protective oxide scales upon exposure to oxidizing environments.

Various commercial processes including electroplating, metalliding, PVD, CVD, etc., have been developed for this purpose. Among these coating techniques, pack cementation, e.g. chromizing or aluminizing of iron-base alloys, receives much attention, especially for fuel processing and energy conversion applications. Among the advantages of pack cementation are: (1) cost effective improvement of the surface stability without expensive equipment, (2) avoidance of annealing post-treatments, (3) applicability to a wide variety of metals, (4) ability to coat pieces of moderate size and different shapes, (5) relative ease to control the coating composition and microstructure.

The cementation pack accomplishes the coating of a substrate by the diffusion of volatile halides of the desired element(s) from the pack to the surface. The pack is composed of a powder mixture of an inert cement (e.g., Al_2O_3), an alloy containing the element(s) to be deposited and a small quantity of a halide salt activator. The active gaseous species are generated by reactions between the activator and the masteralloy in this closed system. Diffusion of the volatile halides of the element(s) takes place between the masteralloy source and the substrate surface, where the activity of the coating element is lower. This activity gradient provides the driving force for the deposition of the coating element.

The reaction step at the surface may be extremely complex, involving adsorption, dissociation, surface diffusion, and a number of other steps. The limiting concentration(s) of the coating element(s) that can be incorporated into the surface of the substrate is set by the thermodynamic limitation that the activity of the coating element cannot exceed that in the source alloy. Moreover, exchange reactions between the reactant gaseous species and the substrate may generate gaseous product molecules of substrate elements for transport into the pack. In other words, the substrate materials could, in a reversal of the coating process, be deposited on the source alloy particles.

Finally, the deposited coating element(s) diffuses into the substrate to form the desired diffusion coating. Solid-state diffusion theory (Fick's second law) applies to this step. The upper limit on the surface composition corresponds to the activity of the coating element in the source alloy. In order to understand the reaction mechanism of a pack cementation process and to control the coating composition and properties, the local thermodynamic equilibria in the pack and at the substrate surface, the gas-solid reactions at the interface, and the kinetics of gas diffusion and solid-state diffusion must be considered [1].

Oxidation protection by coatings is based on the formation, of a thin, compact, adherent, protective oxide scale, either Cr_2O_3, Al_2O_3, or SiO_2, during the early stage of exposure. Singular coatings (aluminizing or chromizing) provide significant oxidation resistance. However, among all the coating compositions, a combination of 4-12 wt% Al in combination with 15-25 wt% Cr at the substrate surface results in better protection for iron than that from Al or Cr alone. In an oxidizing environment, the Fe-Cr-Al alloy produces a thin protective layer of alpha-Al_2O_3 scale at steady-state [2-4]. The relatively high Cr concentration enables the rapid formation of a continuous transient Cr_2O_3 layer to block the oxygen ingress which would otherwise cause the internal oxidation of Al. Since Al is thermodynamically more reactive than Cr, a layer of Al_2O_3 eventually evolves from Al displacement and substitution into the isomorphous Cr_2O_3 scale [5].

The objective of this research was to apply the pack cementation technique to coat iron and iron-base alloys simultaneously with chromium and aluminum to yield a coating layer containing 15-25 wt% Cr and 4-12 wt% Al. The composition range is illustrated in Fig. 1.

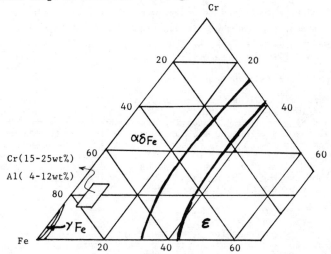

Fig. 1 Partial Ternary Isotherm of Fe-Cr-Al System.

It is well known that the adherence of oxide scales can be significantly enhanced by a small addition of reactive elements such as Zr, Ce, and Y [6-9]. But the mechanism responsible for this enhancement of scale adherence is still controversial. In this study, yttrium was simultaneously doped into a Cr-Al coating on pure iron by directly mixing yttria powder into the cementation pack. The results of thermogravimetic oxidation tests (TGA) and a cyclic oxidation test will be presented.

II. EXPERIMENTAL MATERIALS AND PROCEDURES

1. Specimens and Chemicals

Specimens used in this research included iron of MARZ-grade (99.998% pure) from Morton Thiokol Inc., and Cr-Mo steel (Fe-2.25Cr-1Mo) from Babcock & Wilcox Co. Chemical analysis of the FeCrMo alloy is listed in Table 1.

Table 1. Chemical Analysis of Fe-2.25Cr-1Mo (provided by Babcock and Wilcox)

Element:	C	Mn	Cr	Si	Mo	S	P	Cu	Fe
Conc.(wt%)	.12	.46	2.36	.37	.99	.013	.008	.1	Bal

The masteralloys which provided the deposition sources included a series of 200-mesh Al-Cr alloy powders made by Consolidated Astronautics and Alloy Metals, Inc. Halide activators served to produce volatile halide species for gaseous transport in the pack. In this research, $AlCl_3$, $CrCl_2$, $FeCl_2$, NaCl, and NH_4Cl salts were tested as activators for different coating conditions. The inert cement used in the pack was (80-200 mesh) Al_2O_3 powder. Most of the cementation packs were comprised (by weight) of 25% masteralloy, 2% activator, and 73% inert cement.

2. Experimental Arrangement

The specimen and the mixed masteralloy, activator, and cement powders were packed into a 4 cm OD cylindrical crucible of pure alumina and sealed with a pure alumina lid. The coating experiments were performed in a horizontal resistance-heated tubular furnace with a uniform heated zone of about 10 cm. An assembly was designed to rotate the reaction chamber including the alumina pack at a speed of 10 rpm during the pack cementation process. The intention of this rotation was to homogenize the pack composition throughout the process by the tumbling action, and thereby avoid local alloy depletion near the specimen.

3. Characterization

The surface morphologies of coated substrates were studied with the JEOL-JXA-35 SEM. Concentration profiles were generated by Energy Dispersive Spectroscopy (EDS) interpreted by an EDAX SW-9100 computer system. Element identification and x-ray mapping were performed by Wavelength Dispersive Spectroscopy (WDS). Study of the phase structure was conducted with a Scientag PAD-V x-ray diffractometer.

Isothermal thermogravimetric (TGA) and cyclic oxidation testing in air of as-coated specimens was accomplished using a Cahn R-100 microbalance. Isothermal TGA tests were performed at 1000 C for 40-80 hours. Cyclic oxidation tests were conducted between 50 C and 1000 C. The specimen was held at 1000 C for 10 hours and cooled to 50 C in 30 minutes.

III. RESULTS AND DISCUSSION

1. Cr-Al Coating on Pure Iron

Initially, a masteralloy of Fe-24.0Cr-8.5Al was selected to coat the iron substrate. The tentative assumption was made that the activities of Al and Cr in this pack should be adequate and would be maintained constant by the excess powder. The coating compositions produced by this masteralloy in conjunction with the activators NaCl, NaF, and NH_4Cl are summarized in Table 2.

Table 2. Surface Compositions and Weight Gains for Cr-Al
          Coatings with Fe-24.0Cr-8.5Al Masteralloy,
          Processed at 1000 C for 27 Hours.

Activator	wt% Cr (Surface)	wt% Al (Surface)	Weight Gain (mg/cm^2)
NaCl	1.91	7.47	+3.2
NaF	1.08	9.36	+228.0
NH$_4$Cl	6.67	7.60	-51.0

Regardless of the activator used, the Cr content at the coating surface was
far below the desired value. Significant formation of iron chloride product
vapor resulted when conditions favored Cr deposition. To investigate further
the problem of insufficient Cr deposition, a Fe-19.4Cr specimen was processed
with the same masteralloy and the NaCl activator. After 27 hr at 1000 C, the
Cr concentration at the coating surface dropped to 14%, below the initial Cr
concentration of the substrate, while the surface Al concentration reached
8.5 wt%.

    Marijnissen [10] suggested an important exchange reaction between
aluminum chloride and chromium chloride during co-deposition of chromium and
aluminum on a nickel-base alloy, i.e.,

$$AlCl_2(v) + \underline{Cr} = CrCl_2(v) + \underline{Al} \qquad (1)$$

    According to thermodynamic data, the partial pressures of the
predominant aluminum halides, $AlCl/AlCl_2/AlCl_3$ are several orders of
magnitude higher than the partial pressure for chromium chloride, $CrCl_2$.
Thus, reaction (1) is favored in the forward direction, and $\underline{Cr}$ deposited or
present in the substrate surface may be displaced and replaced by $\underline{Al}$.

    The enhancement of the chromium activity was obtained by adjusting the
Cr content in the masteralloy. Activities for Al in Cr-Al alloys were
measured by Johnson, Komarek and Miller [11]. For a Cr-Al binary masteralloy
of 5wt%Al, $a_{Al} \approx 4 \times 10^{-4}$; for 10wt%Al, $a_{Al} \approx 10^{-3}$ at 1000C. Experiments were
conducted by using various Cr-rich masteralloys in packs activated by NaCl,
NH$_4$Cl, AlCl$_3$, CrCl$_2$, FeCl$_2$, and several combinations of these salts. The
results are summarized in Table 3. Clearly, it was not possible to approach
the desired coating composition, Fe-(15-25)Cr-(4-12)Al, unless the Cr/Al
ratio of the masteralloy was higher than 90/10. Furthermore, the AlCl$_3$
activator was found to enhance greatly the Cr content at the substrate
surface, probably by the action of Reaction (1) in the pack.

Table 3. Effects of Masteralloy and Halide Activator in Rotating Packs Processed at 1000C in Argon for 27 Hours.

MASTERALLOY	ACTIVATOR	Cr wt%	Al wt%
95Cr-5Al	NaCl	13.23	0.08
95Cr-5Al	NH_4Cl	35.90	1.09
95Cr-5Al	$1NaCl:1NH_4Cl$	52.96	1.50
95Cr-5Al	$AlCl_3$	65.35	0.30
95Cr-5Al	$CrCl_2$	36.71	1.75
95Cr-5Al	$1NH_4Cl/1FeCl_2$	30.62	2.82
95Cr-5Al	$1NaCl/1AlCl_3$	17.20	7.31
95Cr-5Al	$1NaCl/2AlCl_3$	20.13	4.02
95Cr-5Al	$1NaCl/3AlCl_3$	14.77	6.68
90Cr-10Al	NaCl	12.28	11.21
90Cr-10Al	$1NaCl:1NH_4Cl$	14.86	13.02
90Cr-10Al	$1NH_4/FeCl_2$	5.69	11.20
85Cr-15Al	NaCl	14.39	15.85
85Cr-15Al	NH_4Cl	5.24	12.80
85Cr-15Al	$1NaCl:1NH_4Cl$	6.32	17.49
80Cr-20Al	$1NaCl:1NH_4Cl$	14.40	20.25
70Cr-30Al	NaCl	6.35	40.2
70Cr-30Al	$1NaCl:1NH_4Cl$	13.50	27.59

Table 3 features the results for a specific testing of mixed $NaCl/AlCl_3$ activator salts with variable ratio for the 95Cr-5Al alloy. The $1NaCl/2AlCl_3$ activator provided the desired Kanthal surface composition.

Figure 2 shows the concentration profiles for this coating run; these profiles which extend to a depth of about 300mm could be closely reproduced. No evidence for small composition discontinuities corresponding to the presence of the gamma phase at low Cr and Al contents was seen; nevertheless, a q/a interface must have existed at 1000C. Cross-sectional metallography showed a virtual absence of any entrapped masteralloy or alumina cement or Kirkendall voids, however some pack particles were imbedded in the immediate surface. Measurements of the specimen dimensions and weight before and after coating showed a negligible change in each parameter. Therefore, exchange reactions between the substrate Fe and $CrCl_2(v)$ and $AlCl(v)$ were in close balance. This condition of constant dimensions is extremely advantageous from a practical engineering standpoint, but would probably be violated in a pack containing H_2 because HCl could serve to remove product chlorine from the substrate.

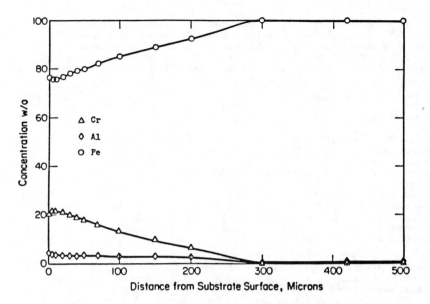

Fig. 2 Concentration profiles of the Cr-Al coating, 27hr, 1000C,
masteralloy: 95Cr-5Al, activator: (2)AlCl₃ + (1)NaCl.

The optimized coating conditions for the simultaneous
chromizing/aluminizing of pure iron are as follows: 95 Cr-5Al masteralloy
(25wt%), 1NaCl/2 AlCl$_3$(2wt%), Al$_2$O$_3$ cement (75%), treated for 27 hours at
1000C in argon. The pack was rotated (tumbled) at 10 rpm. In the absence of
pack rotation, the surface composition became 12.5 Cr and 5 Al(wt%).

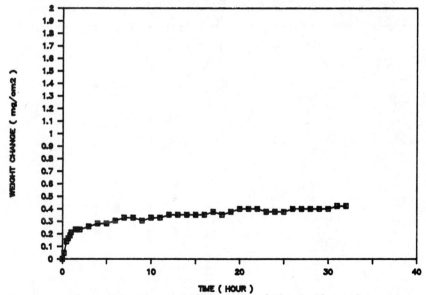

Fig. 3. Oxidation kinetics at 1000C in air of the Cr-Al coating on pure
iron.

137

The as-coated iron specimen with surface composition Fe 20Cr-4Al was oxidized in air isothermally at 1000C in a TGA system. The resulting kinetics, shown in Fig. 3, reveals that the coated iron developed a protective layer of alumina within the first 4 hours and became virtually passive to further oxidation. The weight-gain after 40 hours oxidation was only 0.4 mg/cm^2. Clearly, isothermal oxidation at 1000C is not a severe test for the Kanthal surface composition.

2.   Cr-Al Coating on Fe-2.25Cr-1Mo Steel

The Fe-2.25Cr-1Mo steel (FeCrMo) is commonly used as a structural steel with high temperature applications related to pressure vessels, coal conversion reactors, steam generators, and petrochemical equipment. In this research, a Cr-Al coating process was developed to improve its high-temperature oxidation resistance.

When the same coating process which was so successful for pure Fe was attempted for FeCrMo, X-ray mapping by WDS of the resulting concentration profiles showed a Cr-rich layer about 10mm thick at the surface, and a much lower uptake by both Cr and Al into the substrate. The Cr-rich layer was identified by x-ray diffraction as chromium carbide, $Cr_{23}C_6$, which resulted from reaction with the 0.12 wt% of carbon content in the substrate. Therefore, the chromium activity in the masteralloy was reduced by choosing a 90Cr-10Al composition, and a series of different activator salts, given in Table 4, were tested at 1000C for 27 hours in a rotating pack. The most favorable surface composition resulted from the use of 2wt% of pure $CrCl_2$ activator. The corresponding concentration profiles are shown in Fig. 4.

Table 4. Summary of surface compositions showing activator effects for Cr-Al coating of FeCrMo substrates using 90Cr-10Al masteralloy at 1000C in Ar (pack rotation)

Activator	Cr wt%	Al wt%
NH_4Cl	4.58	11.45
$2AlCl_3/NH_4Cl$	7.9	17
$CrCl_2/AlCl_3$	10.1	18.1
$NaCl/CrCl_2$	12.5	21.4
$CrCl_2$	13.5	7.7

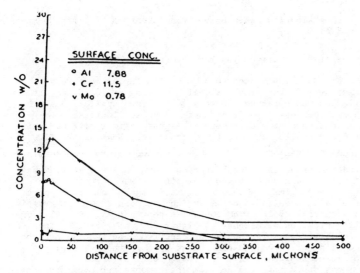

Fig. 4   Concentration profiles for chromized-aluminized FeCrMo alloy using
90Cr-10 Al masteralloy, pure $CrCl_2$ activator, at 1000C in Ar for
27hrs. (Pack was rotated)

According to Perkins [12], an Fe-13Cr-4.5 Al-0.25 Ti alloy coupon
survived the Coal Gasification Atmosphere (CGA) test at 1000F and 1300F for
2000 hours.   Therefore, the Fe-13.5Cr-7.7Al coating produced in this study
should show comparable resistance.   This chromized-aluminized FeCrMo alloy
was oxidized isothermally at 1000C in air for 70hr.   The resulting kinetic
curve is given in Fig. 5.   The kinetics are somewhat higher than the coated
Fe specimen, probably because of the interference of the Mo content.

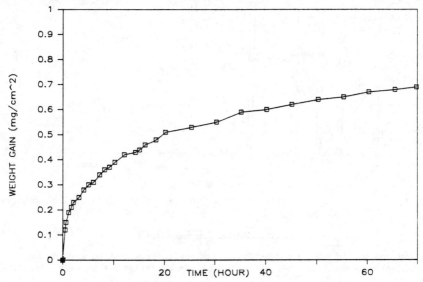

Fig. 5   Isothermal kinetics for chromized-aluminized FeCrMo alloy oxidized
in air at 1000C.

139

Nevertheless, the slow kinetics again assure the formation of a steady-state protective scale of Al_2O_3.

3. Yttrium Doping of Substrate during Pack Cementation Coating

The iron and FeCrMo alloy protected by a Cr-Al coating were very corrosion-resistant during isothermal oxidation testing at 1000C, and no significant spallation was observed. However, localized spallation of the oxide scale occurred upon even the first thermal cycle. Therefore, to enhance the scale adherence, a preliminary attempt to effect yttrium-doping was made for the Cr-Al coating on pure iron substrate. Some of the Al_2O_3 cement was replaced by mixing 2wt% Y_2O_3 submicron powder into the cementation pack. From the rather crude EDAX determination, about 0.4 wt% yttrium was detected at the surface, and some evidence of yttrium was detected to a depth of 80mm. Clearly this preliminary work requires further analytic support. However, the chromized-aluminized specimen which had been doped with yttrium was subjected to 5 thermal cycles from 1000C, as shown in Fig. 6. Although the initial scaling kinetics were somewhat higher for this specimen than for that in Fig. 3, only little scale spallation occurred upon thermal cycling. Similar thermal cycling of chromized-aluminized iron, without Y doping, resulted in significant scale spallation. These results support the conclusion that Y was doped into the Fe from the cementation pack and that the Y improved scale adherence.

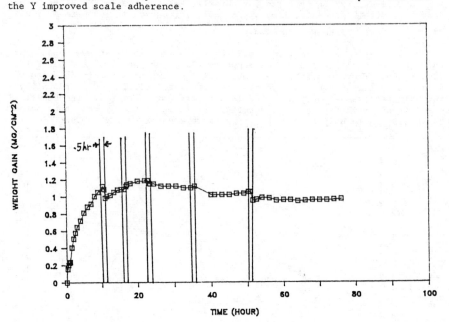

Fig. 6   Cyclic oxidation kinetics at 1000C for a yttrium-doped iron specimen with Cr-Al coating.

IV   SUMMARY AND CONCLUSIONS

By the use of very Cr-rich Cr-Al masteralloy powders in a rotating (tumbling) cementation pack, differing halide activator salts were identified to produce Kanthal-like surface compositions for pure Fe and an Fe-2.25Cr-1Mo alloy. The isothermal oxidation of such coated specimens resulted in slow kinetics, consistent with the formation of an a-Al_2O_3 scale at steady-state. A preliminary attempt to dope yttrium via Y_2O_3 at the same time seemed successful, and resulted in improved cyclic oxidation resistance. A US patent application has been submitted to cover the technology disclosed in this paper.

V ACKNOWLEDGEMENTS

This research was sponsored by Electric Power Research Institute under Project RP-2278-1, monitored by J. Stringer. Helpful collaboration by S.C. Kung is appreciated.

References

1. P.N. Walsh, "Chemical Aspects of Pack Cementation", Chemical Vapor Deposition, J. Blocker, Ed., Electrochem. Society, Princeton, (1973).

2. P. Tamaszewicz, and G.R. Wallwork, "The Oxidation of High-Purity Fe-Cr-Al Alloys at 800C", Oxid. Metals, 20(3/4), (1983) 75-109.

3. J.K. Tien and F.S. Pettit, "Mechanism of Oxide Adherence on Fe-25Cr-4Al(Y or Sc) Alloys". Metall. Trans., 3(6), (1972) 1587-1599.

4. F.A. Golightly, G.C. Wood and F.H. Stott, "The Early Stages of Development of a-Al_2O_3 Scales on Fe-Cr-Al and Fe-Cr-Al and Fe-Cr-Al-Y Alloys at High Temperature", Oxid. Metals, 14(3), (1980) 217-235.

5. E.A. Gulbransen and K.F. Andrew, "Vapor Pressure Studies on Iron and Chromium and Several Alloys of Iron, Chromium, and Aluminum", Trans. Met. Soc. AIME, 221(12), (1961) 1247-1252.

6. M.H. Lagrange, A.M. Huntz, and J.H. Davidson, "The Influence of Y, Zr or Ti Additions on the High Temperature Oxidation Resistance of Fe-Ni-Cr-Al Alloys of Variable Purity", Corr. Sci., 24(7), (1984) 613-627.

7. A.R. Nicoll and G. Wahl, "The Effect of Alloying Additions on M-Cr-Al-Y Systems- An Experimental Study", Thin Solid Films, 95, (1982) 21-34.

8. M. Allam, D.P. Whittle, and J. Stringer, "Improvements in Oxidation Resistance by Dispersed Oxide Addition: Al_2O_3-Forming Alloys", Oxid. Metals, 13(4), (1979) 381-401.

9. T. Amano, S. Yajima and Y. Saito, "High Temperature Oxidation of Ni-20Cr-5Al Alloys with Small Additions of Cerium", Trans. Japan Inst. Metals, 26(6), (1985) 433-443.

10. G.H. Marijnissen, "Codeposition of Chromium and Aluminum During a Pack Process", High-Temperature Protective Coatings, C. Singhal ed., TMS–AIME Pittsburgh, (1980) 27-35.

11. W. Johnson, K. Komarek, and E. Miller, "Thermodynamic Properties of Solid Cr-Al Alloys at 1000C," Trans. Met. Soc. AIME, 242 (1968) 1685-1688.

12. R. Perkins, Lockheed Aerospace Div., Palo Alto, personal communication.

EVALUATION OF SURFACE MODIFICATIONS FOR OXIDATION PROTECTION

OF VANADIUM-BASE ALLOYS

A. G.TOBIN AND G. A. BUSCH

Grumman Corporate Research Center

Bethpage, NY   11714-3580 USA

Abstract

A major drawback in the application of vanadium alloys to the blanket of a fusion reactor is their known sensitivity to oxidation by the impurities in the coolant and their subsequent embrittlement. Oxidation protection of vanadium alloys will be necessary for long-term successful operation at temperatures above 875 K. Surface alloying using chromium was investigated as a means of creating an oxidation-resistant surface. Oxidation tests were then conducted in flowing helium containing up to 100 vppm H_2O impurity. Temperatures up to 925 K and times up to 1000 h were utilized. Control samples of unmodified V-15Cr-5Ti and 316 stainless steel were included. Significant reductions in weight gain and oxygen penetration into the alloy substrate were observed with surface alloyed specimens. Surface alloying thus offers the promise of protecting vanadium alloy surfaces in the impure He environments anticipated in fusion reactor' operation.

143

Introduction

Advanced fusion reactors will need to be constructed of materials that can survive in one of the severest environments ever contemplated for any energy production system. For example, the first wall and blanket materials will be exposed to high-temperature, cyclic thermal stresses, corrosive coolants, high fluxes of 14 MeV neutrons, internal hydrogen and helium generation, and high fluxes of energetic deuterium, tritium, and helium. Although austenitic and ferritic stainless steels remain the current prime candidate alloys for first wall/blanket applications, there still remain many unanswered questions associated with their neutron embrittlement, induced swelling, thermal fatigue behavior, induced activation, and corrosion resistance. As a class, vanadium-base alloys generally appear to have lower swelling after neutron irradiation and retain significant ductility. They also appear to be capable of operating at significantly higher temperatures than stainless steels and exhibit significantly lower activation after neutron irradiation. These unique properties are attractive for fusion reactor blanket design options based upon helium coolants (1). However, a major restriction on the use of vanadium alloys at elevated temperatures is their susceptibility to embrittlement by oxidizing impurities (i.e., H_2O) that are likely to be present in the helium coolant gas.

Although the maximum solubility of oxygen in vanadium is approximately 2 wt% (2), oxygen levels greater than 0.25 wt% are sufficient to embrittle vanadium at ambient temperatures (3). Since a He-cooled vanadium alloy fusion reactor blanket must operate at temperatures up to 700°C for periods exceeding 10^4 h, the identification and control of those factors which influence the diffusion of oxygen from the surface into the bulk alloy will determine the rate of embrittlement of vanadium alloy components.

Analytical studies (4) of the interaction of H_2O with V and V-alloys at elevated temperatures indicated that the presence of moisture levels in excess of 1 vppm H_2O in the coolant gas could lead to nucleation of a V_9O phase below 700°C. Since the expected moisture levels in a He-cooled blanket may reach up to 100 vppm H_2O, it is expected that an oxide will always be present on the surface of vanadium alloy components. Thus, their rate of degradation will be determined by the rate of growth of the oxide and by the rate of oxygen diffusion into the metal in which the flux of oxygen is determined by the oxygen gradient and diffusivity in the metal at the oxide/metal interface. In the case of unalloyed vanadium it was shown (4) that moisture levels below 0.1 vppm H_2O would be needed to avoid total embrittlement within 10^4 h at 775 K. In the case of vanadium alloys, it was not possible to perform a similar assessment since the oxygen diffusivities and oxide-metal equilibria in V-alloys are not known. Yamamoto (5) has shown, however, that considerable improvements in the oxidation resistance of V alloys in air at 975 K may be achieved via additions of Cr and/or Ti to the vanadium alloy matrix. These results suggest that significant improvements in oxidation performance may be expected through suitable alloying of the surface. To examine this approach, V-15Cr-5Ti alloy sheet was selected for surface modification procedures using Cr as the surface alloying element. The selection of Cr was based upon its high solid solubility in V and its known high oxidation resistance. Materials modified in this manner were exposed to flowing He containing 100 vppm H_2O impurities at temperatures up to 925 K and for times up to 1000 h. Comparisons of surface modified alloys with 316 stainless steel and the unmodified alloy were made.

Experimental Procedures

The baseline vanadium alloy, V-15Cr-5Ti (4), was selected for evaluation. The alloy was prepared by Westinghouse via tungsten arc-melting of a 450 g charge of the appropriate quantities of pure elements on a water-cooled Cu hearth. The arc-melted button was canned and subsequently hot-rolled into 0.75 mm thick sheet and cut into 12.5x12.5x0.75 mm test coupons. Energy dispersive analyses (EDS) of the as-received alloys indicated that all the measured compositions were within the nominal limits. To modify the alloy surface, Cr was vapor-deposited on and diffused into the surface at elevated temperatures to form a surface-modified Cr-enriched layer. For comparison purposes, unalloyed V and Type 316 stainless steel also were included in the study. Prior to oxidation all specimens were surrounded by Ti sheet and vacuum annealed at 1275 K for 1 h at 5×10^{-6} torr. This procedure minimized oxygen contamination of the specimens during the vacuum anneal. Prior to oxidation, all specimens were ultrasonically cleaned in acetone and distilled water.

Oxidation experiments were carried out in a stainless steel, three-zone tube furnace assembly consisting of four separately controlled furnaces with individually controlled temperature settings and gas flow meters. Up to 15 test coupons suspended from hooks could be placed into each furnace whose uniform hot zone (±2 K) length was adjusted to 20 cm. Helium gas containing 100 vppm H_2O was used for all exposures. Each furnace was evacuated to a pressure of about 6.65×10^{-4} Pa prior to introduction of the helium at a pressure of slightly above 1 atm. He flow rates in the range of 50 to 100 cm^3/min were utilized in these tests. For these continuous oxidation studies, specimens were exposed in 200 h intervals up to 1000 h and a new set of test coupons was inserted after each exposure. All specimens were weighed before and after oxidation in a microbalance to within an error of ± 5 μg.

Scanning electron microscopy (SEM) was used to characterize the microstructure and scale morphology of the oxides formed. Energy-dispersive X-ray spectrometry (EDS) was used to study corresponding average oxide composition at the corners and at the centers of selected samples. Accelerating voltages of 10, 20 and 30 kV were used to yield X-ray signals that were generated at depths of 1 to 3 μm. This approach permitted the sampling of only the surface oxide, allowed a determination of its type, composition, and yielded an upper limit to its thickness. The elemental distributions in a polished cross-section of the metal substrate were also evaluated using the digital beam scanning capability of the EDS system.

Results and Discussion

Figure 1 compares the effects of oxidation weight gain vs time for V-15Cr-5Ti, 316 stainless steel, and surface-alloyed V-15Cr-5Ti at 925 K. These results show that surface alloying with Cr enrichment can reduce the oxidation rate of the ternary alloy by approximately two orders of magnitude. This rate is comparable to that for 316 stainless steel.

Quantitative EDS analysis of the chemistry of the oxide scale (Table I) indicated that a (V,Ti,Cr)O- type phase formed on the surface of the modified material, whereas a $(V,TiCr)_2O-$ type phase formed on the unmodified materials at 875 K. Based upon counting statistics, we estimate that the error in these measurements is approximately 2%. It was not possible to obtain positive identification of the oxide formed using X-ray diffraction since the diffraction patterns generated did not match any known crystallographic data found in the standard powder diffraction files. Although the identification of the oxide cannot be unambiguously obtained, it is likely that the low oxidizing potential of the moist He (10^{-4}atm H_2O) would lead to the formation of a lower oxide of vanadium. It is also possible that more than one oxide phase could be present in the oxide scale, since the EDS analysis can only provide an average composition in any given depth of the material; however, the closeness of the measured composition to known oxide of V and the uniformity of the oxide scale microstructure suggest that only one oxide phase has indeed nucleated on

146

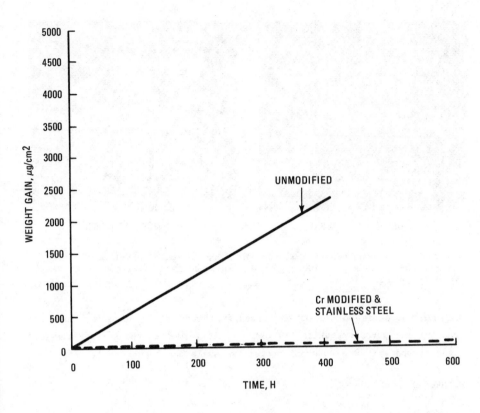

Figure 1 - Effect of Cr surface modification on oxidation resistance of V-15-Cr-5Ti exposed to He containing 100 vppm H_2O at 925 K.

Table I. Composition of Oxide Scale (atom %) of Surface Alloyed V-15Cr-5Ti after Exposure to He Containing 100 vppm H_2O at 875 and 925 K.

Element		875 K/600 h Atom %	925 K/400 h Atom %
Unmodified	V	51.85	44.82
	Cr	10.44	7.18
	Ti	4.49	5.92
	O	33.20	43.79
Modified	V	16.34	13.24
	Cr	21.69	25.12
	Ti	10.11	8.06
	O	51.86	52.31

the surface. The compositions of these oxides are independent of oxidation time. Typical SEM micrographs of the oxide scales formed on V-15Cr-5Ti and surface-modified V-15Cr-5Ti after exposure at 875 K are shown in Fig. 2.

147

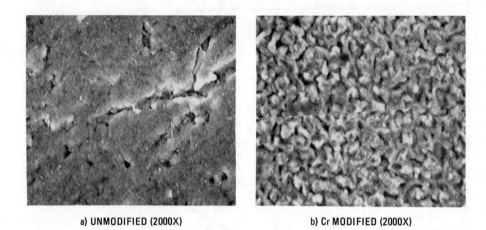

a) UNMODIFIED (2000X)   b) Cr MODIFIED (2000X)

Figure 2 - Scanning electron micrograph of oxide scale in V-15Cr-5Ti alloy exposed to He containing 100 vppm H_2O: a) unmodified after 1000 h at 875 K; b) Cr surface modified.

The surface oxide formed on the unmodified base alloy shows evidence of cracking and porosity, while the oxide scale formed on the surface-alloyed material shows that a coherent protective scale has been formed. Figure 3

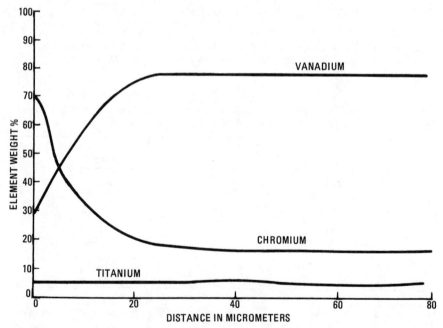

Figure 3 - Elemental distribution in near-surface region of Cr-modified surface in V-15Cr-5Ti.

illustrates the elemental distribution in the near-surface region of a polished cross-section of V-15Cr-5Ti after surface modification by Cr. A Cr-enriched diffusion layer containing up to 70% enrichment in Cr at the surface is observed within the first 20 μm depth. Similarly, Fig. 4 illustrates the corresponding elemental distribution in the near surface region of the Cr-modified material after oxidation at 875 K for 1000 h. The high oxygen level in the first 2 to 3 μm suggests that the material consists primarily of oxide whose exact composition cannot be determined because of the finite width (2 μm) of the scanning electron beam. However, the rapid fall in the oxygen level within the first 4 μm of the surface clearly shows that the oxygen penetration into the alloy substrate has been restricted by the surface treatment. This suggests that the Cr enrichment significantly reduces the oxygen gradient in the near-surface region of the V-Cr-Ti matrix and, thus, the driving force for diffusion in the alloy. This observation is consistent with the known low solubility of oxygen in Cr. The low weight gains observed in combination with the previous observations suggest that most of the oxygen is trapped in the Cr-enriched oxide scale which is both coherent and protective. In contrast, Fig. 5 illustrates the elemental distribution of unmodified V-15Cr-5Ti after

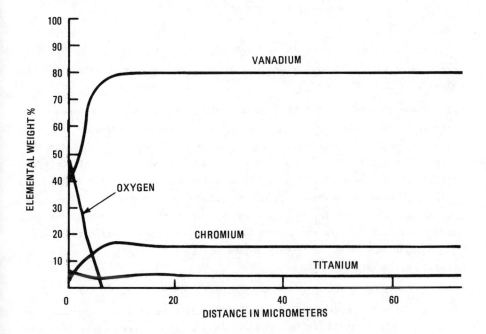

Figure 4 - Elemental distribution in Cr-modified substrate of V-15Cr-5Ti after oxidation in He containing 100 vppm at 875 K for 1000 h.

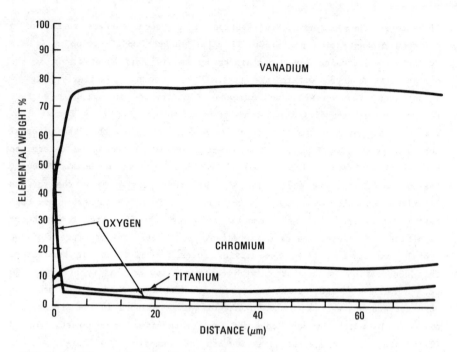

Figure 5 - Elemental distribution in V-15Cr-5Ti substrate after oxidation in He containing 100 vppm H_2O at 875 K for 1000 h.

exposure at 875 K for 1000 h and indicates that significant oxygen penetration of the alloy has occurred at the observed oxygen levels in the alloy substrate.

These results suggest that surface alloying is an effective means of minimizing oxygen embrittlement of vanadium alloys during exposure to moist He at blanket operating temperatures. It is thus clear that the alloying element distribution in the near-surface region has a strong effect upon the nature and the protectiveness of the oxide scale in vanadium alloys, as well as the oxygen diffusion gradient at the oxide/metal interface. In particular, the optimization of elemental distributions in the near-surface region may hold the key to further improvements in the oxidation resistance of vanadium alloys in the impure helium coolants that will be employed in fusion reactor blankets.

Conclusions

1. Oxidation of Cr-modified surfaces up to 925 K in He containing 100 vppm H_2O leads to a significant reduction in oxygen penetration into the base metal and to a more coherent protective oxide film than for corresponding unmodified surfaces.

2. Surface alloying techniques offer great promise for providing oxidation protection of vanadium alloys under He cooled fusion reactor blanket operating conditions.

REFERENCES

1. C.P. Wong et al., The Elongated Tokamak Commercial Reactor Design, 11th Fusion Engineering Symposium, Austin, Texas, Nov. 1985.

2. H.L. Henry, S.A. O'Hare, R.A. McCune, and M.P. Krug, "The Vanadium-Oxygen System; Phase Relations in the Vanadium Rich Regions Below 1200°C," J. Less Common Metals, 21 (1970) 15.

3. B.A. Loomis and O.N. Carlson, "Investigation of the Brittle-Ductile Transition in Vanadium," Reactive Metals 2, Metallurgical Society Conference, New York, NY: Intersicence Publisher,(1959).

4. A. Tobin and J. Bethin, "Blanket Comparison and Selection Study" Argonne National Lab Report No. ANL/FPP-83-1, Oct. 1983.

5. A.S. Yamamoto and W. Rostoker, "Exploration of Vanadium-base Alloys," Wright Air Development Center Technical Report 52-145, Part 2 (1954).

Acknowledgments

The authors wish to acknowledge Ms. Meghan Kennelly for performing the oxidation experiments and W. Poit for performing the surface alloying experiments. This work was performed under internal Grumman funding.

OXIDATION/CORROSION OF SURFACE-MODIFIED ALLOYS

V. Srinivasan

Universal Energy Systems, Inc.
4401 Dayton-Xenia Road
Dayton, Ohio 45432

ABSTRACT

Surface modification by ion beams is a non-equilibrium method of producing novel surface chemistries and microstructures, and therefore alters the surface properties without affecting the bulk properties. This method has been widely applied both to understand the mechanisms of and to control the surface degradation by oxidation/corrosion. The results of several investigations present interesting diversities and also bring out the potential of this method. They are illustrated and discussed in this presentation using the published results and our recent observations on nickel-base alloys.

Introduction

Direct implantation and ion-beam mixing are non-equilibrium methods of modifying the surfaces chemically and microstructurally (1). They are employed to alter the surface properties. The above methods are used to study the mechanisms of oxidation and its control at high temperatures in metals and alloys (2-11). Enduring beneficial effects of implantation have been reported in some alloys even though the initial implanted layer was only a few hundred angstroms in thickness (3,4). Faster oxidation kinetics and scale growth have also resulted from implantation (5,6). It is therefore necessary to reconcile some of the observations and to gain a better understanding of the oxidation/corrosion response of surface-modified alloys in terms of material, implantation, and test variables. This paper discusses some of the above aspects in the background of published information supplemented by our observations.

Implications of Surface-Modification

In direct-implantation as well as in ion-beam mixing high energy ions penetrate the target surface till their kinetic energy falls below a threshold value. They generate along their trajectory physical defects as a result of their collisions with the host atoms. For normal incidence the depthwise distribution of the implants in a polycrystalline substrate will be closely gaussian. This distribution will get modified for other angles of incidence, multiple-energy implants and high sputter yields. The beam energy, dose, sputtering yields, the nature of the ions, and the geometry of implantations determine the depth of penetration, physical damages, and the attainable concentration for a given target. The rate of ion-beam mixing depends on the solubility and the tendency of the deposited atoms to form compounds with substrate atoms, and incomplete ion-beam mixing will leave a layer of deposited metal at the surface. Implantation raises the target temperature. The magnitude of this rise depends on the experimental conditions. It is also possible to raise the target temperature by external heating to promote defect annealing, efficient mixing or compound formation during implantation. For the same alloying species, implantation may lead to a finer distribution of the implant or its compounds without macro-segregation within and/or along the grain boundaries compared to the ingot metallurgy (I/M) route. Thus ion-implantation forms a distinct method of incorporating elements in the surface of a substrate resulting in a thin modified-surface layer. The physical defects generated by the process and the chemical nature of implants may affect the subsequent oxidation/corrosion behavior of the substrate transiently or enduringly. Consequently, the oxidation/corrosion response of surface-modified alloys could be significantly different from that of I/M alloys of identical bulk composition. Typical physical effects of 2 MeV Pt ion implantation in IN939 nickel-base superalloy is shown in Figures 1(a-d) (6). The well-defined spherical γ' precipitates (Figure 1(a)) were disrupted and high density of dislocation loops (Figure 1(b)) was generated by the implantation process. The disruption of precipitates was further noticed in the absence of superlattice reflections from the implanted alloys (Figures 1(c,d)) (6).

High Temperature Oxidation

Ni- and Fe-based alloys have been the preferred substrates in the oxidation study of implanted systems. Implants were mostly oxygen active elements, noble metals, or elements that form protective oxide scale by selective oxidation. Added in bulk these elements and stable oxide dispersions improve the oxidation resistance by reducing the kinetics

154

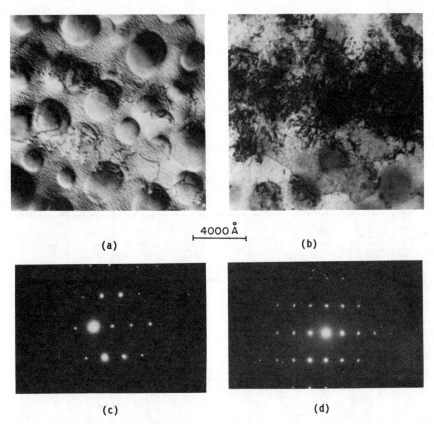

4000 Å

(a)   (b)

(c)   (d)

Figure 1   (a) TEM of unimplanted IN939 alloy. Dark field image by the
           matrix diffraction spot 200. (b) Bright field TEM of Pt
           implanted IN939. (c) TED of unimplanted IN939. Zone axis
           [013]. (d) TED of Pt implanted IN939. Zone axis [$\bar{1}$12].
           Note the absence of superlattice reflections (Ref. 6).

and/or enhancing the scale adhesion (12). The primary objective of these
investigations was to examine and establish the merits of surface
modification procedure in oxidation studies over the I/M route as this
procedure has several advantages such as letting the bulk mechanical
properties remain unaltered. Table I gives the details on the
substrates, implants, and implantation and test conditions along with the
references.

    The effect of implantation had been studied in binary, ternary, and
commercial alloys (2-11). Y-implantation had only a marginal beneficial
effect on the isothermal oxidation behavior of René N4 single crystals at
1000°C in flowing air at 1 atm (7). Both the implanted and unimplanted
alloys developed multilayered scales at 1000°C, Figures 2-4. The outer
layer was rich in Cr and Ti. Below this was a chromia layer. A
protective alumina scale rich in refractory elements was formed at the
metal/scale interface. The metal/scale interface in an implanted alloy
was relatively regular and smooth, Figure 4(a) while it was rough and
irregular, Figures 4(b), in an unimplanted one.

TABLE I
EXPERIMENTAL DETAILS AND REFERENCES

Alloy	Type	Implantation Condition					Oxidation Condition			Ref.
		Energy KeV	Dose cm^{-2}	Surface Peak Conc wt %	Depth, Å		Test Temp °C	Test Duration Hours	Environment	
					Peak Conc	Implantation				
Ni-33Cr	Y	150	2×10^{14}	0.134	290	--	900	40	O_2 1 atm	9
	Y	190	9×10^{15}	2.68	720	--	900	40	O_2 1 atm	9
	Ar	25	1×10^{15}	0.71	100	--	900	40	O_2 1 atm	9
Ni-20Cr	Cr	170	2×10^{16}	3.49	--	1000	1000	24	O_2 1 atm	10
	Al	100	2×10^{16}	0.94	--	1000	1000	24	O_2 1 atm	10
	Ca	140	2×10^{16}	2.08	--	1000	1000	24	O_2 1 atm	10
	Zr	290	2×10^{16}	10.76	--	1000	1000	24	O_2 1 atm	10

TABLE I
EXPERIMENTAL DETAILS AND REFERENCES (CONT'D)

Alloy	Implantation Condition						Oxidation Condition			
	Type	Energy KeV	Dose cm^{-2}	Surface Peak Conc wt %	Depth, Å Peak Conc	Depth, Å Implantation	Test Temp °C	Test Duration Hours	Environment	Ref.
Ni-20Cr	Y	280	2×10^{16}	10.22	--	1000	1000	24	O_2 1 atm	10
	Ce	250	2×10^{16}	25.37	--	1000	1000	24	O_2 1 atm	10
IN939	Y	180	1×10^{15}	0.45	378	--	950	50	With and Without Na_2SO_4 Coating, 1mg/cm^2 1 atm	8
	Ce	2000	8×10^{14}	0.21	2400	--	950	50	With and Without Na_2SO_4 Coating, 1mg/cm^2 1 atm	8
	Pt	2000	2.6×10^{16}	15.95	2028	--	950	50	With and Without Na_2SO_4 Coating, 1mg/cm^2 1 atm	8

TABLE I
EXPERIMENTAL DETAILS AND REFERENCES (CONT'D)

| Alloy | Type | Implantation Condition | | | | | Oxidation Condition | | | |
		Energy KeV	Dose cm^{-2}	Surface Peak Conc wt %	Depth, Å Peak Conc	Depth, Å Implantation	Test Temp °C	Test Duration Hours	Environment	Ref.
René N4	Y	180	3×10^{15}	0.43	378	--	1000	400	Air 1 atm	7
Fe-20Cr-25Ni/Nb	Ce	300	10^{16}	3.2	500	1000	825	5900	CO_2 1 atm	4
	Ce	300	10^{17}	32.0	500	1000	825	5900	CO_2 1 atm	4
	Pt	300	10^{16}	4.2	380	800	825	3000	CO_2 1 atm	4
	Pt	300	10^{17}	40.0	380	800	825	3000	CO_2 1 atm	4
	Al	200	10^{16}	2.5	1500	2700	825	3000	CO_2 1 atm	4
	Al	200	10^{17}	12.5	1500	2700	825	3045	CO_2 1 atm	4

TABLE I
EXPERIMENTAL DETAILS AND REFERENCES (CONT'D)

| Alloy | Implantation Condition | | | | | Oxidation Condition | | | | |
| | Type | Energy KeV | Dose cm^{-2} | Surface Peak Conc wt % | Depth, Å | | Test Temp $^{\circ}C$ | Test Duration Hours | Environ-ment | Ref. |
					Peak Conc	Implan-tation				
Fe–20Cr–25Ni/Nb	Y	300	3×10^{15}	0.65	650	1500	700–850	5000	CO_2 1 atm	11
			10^{17}	21.0	650	1500	825	7157	CO_2 1 atm	14
	Si	250	7×10^{16}	7.75	1500	3000	825	3544	CO_2 1 atm	14
	Ni	300	10^{17}	12.0	800	1800	825	7157	CO_2 1 atm	14
Fe–15Cr–4Al	Y + Al	300	3×10^{15}	0.6	600	1300	1100	3271	Air 1 atm	5
		200	10^{17}	12	1300	2500			Air 1 atm	
	Al	200	2×10^{16}	2.4	1300	2500	1100	3271	Air 1 atm	5

159

TABLE I
EXPERIMENTAL DETAILS AND REFERENCES (CONT'D)

| Alloy | Implantation Condition | | | | Depth, Å | | Oxidation Condition | | | |
	Type	Energy KeV	Dose cm^{-2}	Surface Peak Conc wt %	Peak Conc	Implantation	Test Temp °C	Test Duration Hours	Environment	Ref.
Fe–25Cr– 1.5Al–0.1Y	Al	250	2x10^{16}	--	--	--	1100	24	Air 1 atm	16
			5x10^{16}	--	--	--	1100	24		
			1x10^{17}	--	--	--	1100	24		

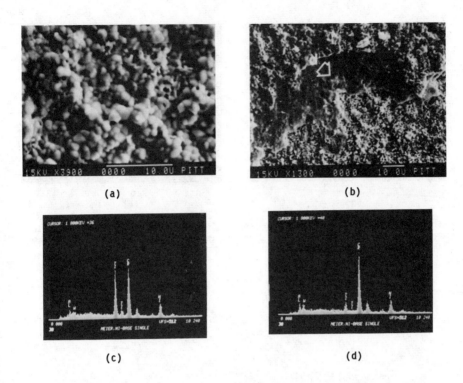

(a)   (b)

(c)   (d)

Figure 2   SEM of (a) the continuous oxide scale, (b) spalled area
(arrow), (c) the EDAX spectra from the continuous scale and
(d) the spalled area on unimplanted René N4 oxidized at 1000°C
for 150 h in flowing air at 1 atm.

(a)   (b)

Figure 3   (a,b) SEM of the oxide scale on the Y implanted René N4
oxidized at 1000°C for 150 h in flowing air at 1 atm.

Implantation of Y, Ce or Pt resulted in higher weight gains and
faster kinetics in another commercial nickel-base alloy, IN939, at 950°C
in flowing air at 1 atm, Figure 5 (8).   This alloy forms a steady state
protective chromia scale below the transient oxides.   René N4 and IN939

161

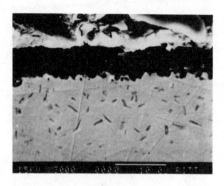
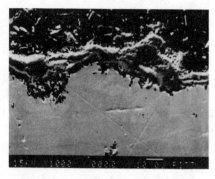

(a)   (b)

Figure 4   SEM of the cross sections of (a) Y implanted René N4 after
150 h oxidation at 1000°C in flowing air at 1 atm, and
(b) unimplanted.

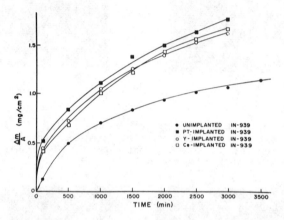

Figure 5   Thermogravimetric data on unimplanted and Pt, Y, or Ce
implanted IN939 oxidized at 950°C for up to 60 h in flowing air
at 1 atm.

are both nickel-base alloys strengthened by γ' precipitates.   Their
nominal chemical compositions are given in Table II.   The size, shape,
distribution, and the volume fraction of γ' precipitates are different
in both the alloys.

Ni-33 Cr and Ni-20 Cr alloys exhibited slower oxidation kinetics and
reduced weight gains after Y-implantation (9,10).   The test conditions
were 40 h 900°C and 24 h 1000°C.   Either Ar implantation in Ni-33 Cr or
Cr implantation in Ni-20 Cr had no effect on the oxidation kinetics of
either of the alloys at the same test conditions (9,10).   The improvement
in the oxidation resistance of Y-implanted Ni-33 Cr was dependent on the
Y-dose.   The parabolic rate constant fell from 6×10^{-7} to
2×10^{-7} mg^2 cm^{-4} S^{-1} when the Y dose was increased from 2×10^{14}
to 9×10^{15} ions/cm^2 (9).   In unimplanted alloy, the outer scale was
essentially NiO with a small percentage of Cr_2O_3 after about 40 h.

TABLE II
NOMINAL CHEMICAL COMPOSITIONS, WT.%

	René N4	IN939
Cr	9.25	22.52
Mo	1.5	0.05
Al	3.7	1.96
Ti	4.5	3.69
W	6.0	2.06
Co	7.0	19.10
Ta	4.0	1.38
Cb	0.5	0.99
C	--	0.14
Zr	--	0.06
Cu	--	0.05
Ni	Bal	Bal

The low dose did not eliminate this outer transient oxide layer, and Y was detected in the inner Cr_2O_3 scale with a peak close to the Cr_2O_3/metal interface. In the case of high dose implantation, the outer transient layer was absent in the sputter depth profiling by SIMS. Further, the Y distribution across the Cr_2O_3 scale was almost uniform dropping sharply near the scale/metal interface (9).

Among Fe-based alloys Fe-20 Cr-25 Ni/Nb stabilized stainless steel has received greatest attention, and formed the substrate for extensive studies to understand the role of several implants in its oxidation behavior in CO_2 or oxygen at temperatures 700-850°C (4,13-15). The protective oxide layer in this alloy was chromia. The implants were Al, Si, Pt, Ce, and Y. The effects of conjoint implantation of elements such as Al, Ce, and Y, and the influence of dose were studied. The experimental details are given in Table I. Al, Si, Pt, and Ni implantations did not alter the oxidation kinetics and weight gains. Even the conjoint implantation of Al and Y did not improve the scale adhesion. Ce and Y implants improved the oxidation resistance when they were implanted individually or conjointly above certain threshold levels (4,14). The thresholds were 10^{16}/cm^2 for Ce and between 5×10^{15} and 5×10^{16}/cm^2 for Y.

The beneficial effects of Al implantation was reported for Fe-25 Cr-1.4 Al-0.1 Y alloy (16). The parabolic rate constants for 24 hr tests in air at 1100°C decreased significantly with increasing Al dose. Such beneficial effects of Al implantation were absent in Fe-15 Cr-4 Al. When Al and Y were implanted conjointly, the protection lasted for only 784 hrs in air at 1100°C (5).

High Temperature/Hot Corrosion

Little is known about the effect of minor additions of rare earths, noble metals or stable oxide dispersion by bulk incorporation or implantation on high temperature multioxidant or deposit-induced corrosion. Improvement in the integrity and adhesion of protective oxide scales by the above additions is expected to extend the incubation period of corrosion thereby delaying the breakaway stage. Thermogravimetric measurements were made on Na_2SO_4 coated coupons of Y, Ce or Pt implanted IN939 alloy. They showed extended incubation stage (8). These observations were made in tests of duration up to 150 hrs at 950°C. Table I describes experimental details. Further reduced internal sulfidation was noticed, Figure 6, in Y-implanted alloy.

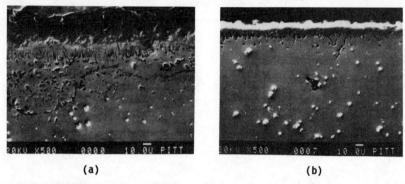

(a)   (b)

Figure 6   Cross-sectional SEM of IN939 after hot corrosion at 950°C.
(a) unimplanted coupon for about 160 h and (b) Y implanted coupon for about 90 h.

Discussion and Conclusions

It is thus evident that the implantation of reactive elements did not improve the oxidation resistance in some alloys while their bulk incorporation in such alloys did (5). This raises an important point as to the usefulness and limitations in the application of ion-implantation process. It is therefore necessary to understand the reasons for the lack of improvement as well as for the reported improvement in the long term oxidation resistance of several alloys by reactive elements implantation. Not all the mechanisms of reduced kinetics and enhanced scale adherence speculated on the basis of observations on the alloys with the bulk additions of reactive elements may be operative in the implanted alloys. Particularly, the mechanisms of the oxidized intermetallic compounds along the grain boundaries acting as pegs may not be valid in implanted alloys because of the possible fine dispersion of implants or their compounds and the consequent absence of macrosegregation along the substrate grain boundaries.

Test duration appears to be an important factor because of shallow depth of incorporation, and need be considered besides other oxidation conditions in assessing and comparing the observations on the implanted surfaces with those on I/M alloys. In comparable test durations, the oxidation kinetics was significantly slower in Y and Ce implanted Ni-33 Cr and Ni-20 Cr, but faster in Y and Ce implanted IN939 compared to that of the corresponding unimplanted system (8-10). Further, the above observations in the implanted Ni-Cr binary alloys are similar to those made in Ni-Cr systems with bulk additions of Y and Ce (17). Ecer and Meier reported a decrease in the oxidation kinetics of Ni-49 Cr containing 0.004, 0.01, 0.03, and 0.08 Ce in the 800-1100°C range (17). The above results were from 24-hr tests and showed a considerable decrease in the parabolic rate constant with increasing Ce concentration. The binary alloys, Ni-33 Cr and Ni-20 Cr, and the commercial IN939 develop protective chromia scales at 800-1100°C by outward cation movement. The peak concentrations of Y in Ni-33 Cr and IN939 were of the same order. It is not therefore clear why Y (and Ce) implantation enhanced the kinetics in IN939 contrary to the observations in Ni-Cr binary systems. It is interesting to note that the kinetics of oxidation in implanted IN939 was almost same within the experimental scatter regardless of the type of implants and the energy of implantation. The only common parameter appears to be the energy deposited by the implantation process. The computer simulation of Y, Ce or Pt implantation process in IN939 at 180 or 2000 KeV gave similar values for the total radiation damage. If the enhanced oxidation kinetics in the implanted IN939 was due to the physical effects of irradiation, the influence of the latter was absent in Ar implanted Ni-33 Cr and Cr implanted Ni-20 Cr as indicated by the unaltered kinetics (9,14). It is speculated that irradiation effects in the form of disrupted γ' precipitates and the associated dislocation tangles might be responsible for the observed faster kinetics in IN939.

Both the implantation and bulk incorporation of Y and Ce have led to similar improvements in the oxidation resistance of Fe-20Cr-25Ni/Nb stabilized austenitic steel (4,15). However, significantly higher concentrations of Y and Ce were required above certain threshold values to achieve the improvement by implantation process. Y and Ce were present across the scale. Y remained mostly in the metal/scale interface while Ce exhibited a peak in Mn-rich scale outer scale (4,9). It is important to note that implanted Ce was detected across the scale width significantly greater than the initial depth of implantation. This suggests redistribution of Ce during the scale growth. On the basis of XRD, SIMS, and ESCA results, Bennett and Co-workers concluded that the fine oxide grains and the presence of CeO_2 along the oxide grain boundaries in Ce implanted Fe-20Cr-26Ni/Nb were responsible for the improved oxidation resistance (4,13,14). The CeO_2 in the grain boundaries reduced the reactant transports along the grain boundaries while small grains contained almost negligible short circuit paths for the outward cation movement. Similar mechanisms were proposed to explain improvements by other rare earth elements.

It is necessary to have implants before the formation of protective oxides for improvement. The implantation of Ce into preformed oxide scale did not improve the oxidation resistance and scale spallation in subsequent tests (14). The initial nucleation and growth process and the reactive element effect thereon appear to be important in determining the long term oxidation behavior. This implies that the various steps such as the oxidation of implants in the preoxidized scale, the grain boundary segregation of the oxidized implants and the subsequent formation of fine-grained oxides are either absent or ineffective.

The absence of any improvement in the oxidation resistance of Y implanted Fe-15Cr-4Al was attributed to the loss of Y with the spalled alumina scale (5). This alloy is an alumina former, and the alumina scale in general forms by the inward movement of anion. Hence, once the growing scale incorporated the implanted Y, the subsequent scale spallation removed Y thereby reverting the oxidation process to that of the unimplanted substrate.

Implantation of Al into the surface of Fe-20Cr-25Ni/Nb stabilized austenitic steel did not lead to the formation of alumina scale. Al apparently got incorporated into the Cr_2O_3 scale probably as oxide particles. This behavior appears to be due to very slow Al diffusion in the austenitic matrix (18). Pt was another element that did not affect the oxidation behavior of the chromia forming F-20Cr-25Ni/Nb. This observation is in contrast to the beneficial effects of Pt reported in several alumina formers (19).

One of the obvious ways by which reactive element additions can affect the high temperature corrosion process is through enhancing the scale adhesion and plasticity. In the absence of scale spallation and microcracks, the corrosive species need be transported by dissolution and diffusion through the oxide scale (20). Extended incubation stage reported in implanted IN939 is believed to be due to more adherent scale. Initial kinetics of hot corrosion as measured by thermogravimetric method was also higher in implanted alloys. Further work to understand the location and chemical status of the implants in corrosion scales is underway. It remains to be established whether the reduced internal sulfidation in Y implanted IN939 was the result of reduced kinetics or chemical effect.

It appears that implantation will in general be effective in alloys that develop scales by the outward movement of cation as in the chromia formers. In some cases such as IN939, the physical damage of implantation may override the chemical influence. Currently, the factors that are responsible for the decisive role of physical defects in some alloys are not well understood. Further, the effect of long term thermal cycling on the efficacy of implantation even in chromia formers such as Fe-20Cr-25Ni/Nb has not been established. Implantation will not be effective in long term tests or service when the scale growth occurs by the inward anion movement as in alumina formation. Research is required to understand the role of implants in alloys that form stratified scales. To be effective the implants must be present before the nucleation of protective oxides. Their effects on the oxide nucleation in chromia formers appears to be a deciding factor, and not their subsequent incorporation in the preformed scale in imparting long term benefits. Further, the effectiveness of the implants depends on the substrate chemistry, microstructure and the nature of the protective oxides that form by selective oxidation. Several mechanisms such as modification of nucleation process and growth kinetics by blocking the diffusion along the grain boundary, making the lattice diffusion difficult, providing vacancy sinks, formation of graded seal, improvement in the oxide-metal chemical bond and enhancing the scale plasticity are likely to be responsible for the observed improvement in ion implanted chromia formers. To identify and understand the operative mechanism in oxidation and corrosion, it is necessary to know the initial and subsequent location and the chemical status of the implants, and the composition and the morphology of the scale. Surface analytical techniques and cross-sectional TEM will help achieve this goal, and their use is strongly recommended.

Acknowledgments

This paper was prepared as part of the research program funded by the Advanced Research and Technology Division, Martin Marietta Energy Systems, Inc. under contract 86X-95901C. The Program Manager is Mr. R. R. Judkins.

References

1. J. K. Hirvonen, ed. "Ion-Implantation" in Treatise on Materials Science and Technology, Vol. 18 New York, NY: Academic Press, (1980).

2. A. Galerie, M. Caillet, and M. Pons, "Oxidation of Ion Implanted Metals," Materials Science and Engineering, 69 (1985) 329-340.

3. M. J. Bennett, "The Role of Ion Implantation in High Temperature Oxidation Studies," in High Temperature Corrosion, ed. R. A. Rapp, NACE, Houston, TX (1983), 145-154.

4. M. J. Bennett, G. Dearnaley, M. R. G. Houlton, R. W. M. Hawes, P. D. Goode, and M. A. Wilkins, "The Influence of Surface Ion Implantation Upon the Oxidation Behavior of A 20%Cr-25%Ni, Niobium Stabilized Austenitic Stainless Steel, in Carbon Dioxide, at 825°C," Corrosion Science, 20 (1980), 73-89.

5. M. J. Bennett, M. R. Houlton, and G. Dearnaley, "The Influence of the Surface Ion Implantation of Aluminum and Yttrium Upon the Oxidation Behavior of A Fe-15% Cr-4% Al Fecralloy Stainless Steel, in Air, at 1100°C," Corrosion Science, 20 (1980), 69-72.

6. V. Srinivasan, G. H. Meier, A. W. McCormick, and A. K. Rai, "Ion Implantation and Thermal Oxidation," Nuclear Instruments and Methods in Physical Research, B16 (1986), 293-300.

7. V. Srinivasan, Development of Surface Modification Method by Ion Beam for Enhanced Oxidation Resistance, Final Report, NSF-DMR-8460741 (1985).

8. V. Srinivasan, Development of Hot Corrosion-Resistant Alloys Using Ion Beam Processing, Final Report, DOE-DE-AC01-83ER80044 (1984).

9. J. C. Pivin, C. Roques-Carmes, J. Chaumont, and H. Bernas, "The Influence of Yttrium Implantation on the Oxidation Behavior of 67 Ni-33 Cr, Fe-43 Ni-27 Cr and Fe-41 Ni-25 Cr-10 Al Refractory Alloys," Corrosion Science, 20 (1980), 947-962.

10. F. H. Stott, J. S. Punni, G. C. Wood, and G. Dearnaley, "The Influence of Ion Implantation on the Development of Cr_2O_3 Scales on Ni-20% Cr at High Temperature" in Ion Implantation into Metals, eds. V. Ashworth, W. A. Grant, and R. P. M. Procter, New York, NY: Pergamon Press, (1982), 245-254.

11. J. E. Antill, et al., "The Effect of Surface Implantation of Yttrium and Cerium Upon the Oxidation Behavior of Stainless Steels and Aluminized Coatings at High Temperatures," Corrosion Science, 16 (1976), 729-744.

12. D. P. Whittle and J. Stringer, "Improvement in Properties: Additives in Oxidation Resistance," Phil. Trans. R. Soc. Lond., A295 (1980), 309-329.

13. M. J. Bennett, B. A. Bellamy, C. F. Knights, and Nicola Meadows, "Improvement by Cerium and Yttrium Ion Implantation of the Oxidation Behavior of a 20 Cr-25 Ni Niobium-Stabilized Stainless Steel in CO_2," Materials Science and Engineering, 69 (1985), 359-373.

14 M. J. Bennett, G. Dearnaley, M. R. Houlton, and R. W. M. Hawes, "The Influence of Cerium and Yttrium Ion Implantation upon the Oxidation Behavior of a 20% Cr/25% Ni/Nb Stabilized Stainless Steel in Carbon Dioxide at 825°C" in Ion Implantation Into Metals, eds. V. Ashworth, W. A. Grant, and R. P. M. Proctor, Pergamon Press, New York, NY (1982) 264-276.

15. M. J. Bennett, H. E. Bishop, P. Chalker, and A. T. Tuson, <u>The Influence of Cerium, Yttrium and Lanthanum Ion Implantation Upon the Oxidation Behavior of a 20% Cr/25% Ni/Nb Stainless Steel in Carbon Dioxide at 900-1050°C</u>, Paper presented at the Surface Modification by Ion Beam Meeting at Kingston, Ontario, Canada, July 1986.

16. U. Bernabai, M. Cavallini, G. Bombara, G. Dearnaley, and M. A. Wilkins, "The Effects of Heat Treatment and Implantation of Aluminum on the Oxidation Resistance of Fe-Cr-Al-Y Alloys," <u>Corrosion Science</u>, 20 (1980) 19-25.

17. G. M. Ecer and G. H. Meier, "The Effect of Cerium on the Oxidation of Ni-50 Cr Alloys," <u>Oxidation of Metals</u>, 13 (1979), 159-180.

18. N. V. Bangaru and R. C. Krutenat, "Diffusion Coatings of Steels: Formation Mechanism and Microstructure of Aluminized Heat-Resistant Stainless Steels," <u>Journal of Vacuum Science and Technology</u>, B2 (1984), 806-815.

19. E. J. Felten and F. S. Pettit, "Development, Growth, and Adhesion of Al_2O_3 on Platinum-Aluminum Alloys," <u>Oxidation of Metals</u>, 10 (1976), 189-223.

20. P. Singh and N. Birks, "Penetration of Sulfur Through Preformed Protective Oxide Scales," <u>Oxidation of Metals</u>, 19 (1983), 37-48.

THE USE OF A CHROMIZED COATING TO PREVENT STRESS CORROSION CRACKING IN AN AMMONIA ENVIRONMENT

W. L. Wentland, W. J. Durako

G. W. Goward

Sundstrand Aviation
4747 Harrison Ave.
Rockford, IL 61125

Turbine Components Corp.
Box 431, Commercial Street
Branford, CT 06405

ABSTRACT

A chromized diffusion coating was evaluated and chosen to prevent stress corrosion cracking (SCC) in a fuel tube. A failure had occurred in the hydrazine fuel tube, and the analysis showed that it exhibited characteristics of stress corrosion cracking. Subsequent SCC testing demonstrated cracking tendencies in ammonia vapor, one of the substances present in the fuel tube during the periods of nonoperation. A chromized coating was chosen to prevent SCC rather than changing to a new base alloy. While chromized coatings are readily used for jet engine applications, this unique application required an extensive development effort. The critical surface to be coated in an ID surface, 0.090″ in diameter and over 3″ in length. Also, the base metal, Hastelloy B, is a solid solution nickel alloy and behaves quite differently from most jet engine alloys. This chromized coating was developed and successfully applied to the fuel tube. Characterization of the coating structure and behavior was also performed. Finally, the coating was tested and found to prevent stress corrosion cracking.

INTRODUCTION

A failure occurred in a hydrazine fuel tube and was attributed to stress corrosion cracking (SCC). The environment responsible for SCC was the decomposition products of the hydrazine (N_2H_4) propellant, which are primarily ammonia and water vapor. The Hastelloy B fuel tube was exposed to this environment during long periods of nonoperational storage. No simple redesign of the tube was considered practical, so the novel approach of using a chromized coating was chosen to prevent SCC in future hardware. This approach allowed for enhanced corrosion resistance at the material surface without altering the intrinsic characteristics of the base metal.

Although chromizing is widely used to enhance the oxidation resistance in jet engine applications, this application provided several unique challenges for the chromized coating. First, the critical surface of the fuel tube to be coated was an ID surface over 3″ in length with a diameter of 0.090″. The coating needs to be uniform and reproducible over this surface. Second, the Hastelloy B base metal (Table A) is a

Table A Chemical Composition and Properties of Hastelloy B

Element	Nominal Percent
Mo	26 to 30
Fe	4 to 6
Co	2.5 max.
Cr	1.0 max.
Si	1.0 max.
Mn	1.0 max.
C	0.05 max.
V	0.2 to 0.4
P	0.025 max.
S	0.030 max.
Ni	Balance

Property	Typical
Ultimate Tensile Strength	115 ksi
Tensile Yield Strength	52 ksi
Tensile Elongation	68%

solid solution nickel alloy with no chromium, which behaves differently from most jet engines alloys that are chromized. Finally, the purpose of the chromized coating was to prevent SCC and was different from conventional use of chromizing.

COATING PROCESS DEVELOPMENT

The basic coating precept was to provide a uniform repeatable coating along the entire tube ID with a thickness of 1.0 mil and a minimum of 25 percent chromium at the surface. Based on prior experience gained in chromizing gas turbine components, the three following variations of existing production processes were considered for coating the Hastelloy B fuel tube:

1. Static gas phase (commonly referred to as "out-of-pack" or "non-contact") wherein the part is suspended above the source of coating species.

2. Dynamic or flowing gas phase: wherein the part is subject to a forced flow of the coating species with an argon carrier gas.

3. Pack cementation: wherein the articles are filled and suspended by the coating species in powder form.

Potential sources of the coating species included chromium powder diluted with aluminum oxide to prevent sintering, and chromium granules with no dilution. In each case, a gaseous carrier species is generated by the reaction of chromium with a halide (chloride or flouride) activator. Also, all coating was done in suitably sized Inconel boxes, closed with channel seals, and contained in retorts through which high-purity argon was flowed. Typical examples of results obtained from the preliminary coating trials are given in Table B. Conclusions were as follows:

- Under the time and temperature conditions practical for the base metal, the static gas phase technique did not yield adequate coating thickness on internal tube surfaces.

- Forced flow through the tube provided no improvement. Theoretically, this method should have provided better results, but the scale of the equipment available did not allow refinement of the process.

- Pack cementation provided a satisfactory coating on all internal tube surfaces. External surfaces were also coated. No difficulty was found in filling the tubes or removing the mix after the time-temperature cycle.

Table B Typical Results of the Experimental Coating Processes Run at 1940°F for 7.75 Hours

Process Type	Chromium Source	Coating Thickness (mils)	Coating Integrity
Static Argon	Cr Powder Cl⁻ Activator Al_2O_3 Powder	0.0-1.0	Spotty, Irregular External Only
Dynamic Argon	Cr Granules F⁻ Activator	0.2-1.0	Spotty, Irregular
Pack Cementation	Cr Powder F⁻ Activator Al_2O_3 Powder	1.1-1.3	Continuous, Uniform on All Surfaces

A pack mix containing 15 percent chromium powder (-325 mesh), 0.5 percent flouride activator, and the balance calcined aluminum oxide (-100 mesh) was chosen for the remaining experimental and production qualification trials. The coating thickness specification was selected to be 0.8 to 1.5 mils, with a chromium content to be a minimum of 25 percent in the outer 15 percent of the coating.

The program then moved towards establishing a frozen process with a rigorous and unique quality control plan to provide the highest possible assurance of obtaining satisfactory coatings on all production pieces. The quality control measures are as follows:

1. A batch of mix is prepared by careful weighing of the specified quantities of chromium, activator, and aluminum oxide. This mix is then blended in a stainless steel V-shell blender and stored in a stainless steel drum under a continuously maintained argon atmosphere. A sample is analyzed for chromium and flouride contents.

2. A pack mix qualification run is made on a Hastelloy B sample. This piece is sectioned for thickness measurements by optical microscopy and for chromium percent by electron beam microprobe analysis. If all analyses showed that the designated mix yields a coating that meets specification limits, then it is designated to be a qualified mix, and remains so for 21 days.

3. Production pieces, each transported in a case designed to minimize transportation damage, are cleaned in acetone to remove handling residues. No other surface preparation is required, and the parts are then handled only with lint-free white gloves.

4. The parts are suspended in a vertical position in a small amount of powder mix in the coating container and manually filled with mix. Light tapping is allowed to minimize void formation. The parts are then completely covered with mix, some mix is placed in the seal, and the cover containing a thermocouple well is placed on the box and sealed (Figure 1).

5. With the thermocouple in place, the container is positioned on a rack and placed in a retort. This retort is sealed, purged with argon, and inserted into the gas-fired furnance. The parts are brought to temperature and the cycle begins.

6. Upon completion of the thermal cycle, the retort is removed, the box is unsealed, and the free-flowing powder mix is removed from the parts. The parts are washed with demineralized water before final inspection. Inspection includes analysis of water used to leach the parts for critical traces of species from the pack mix, such as flouride ions.

7. A test piece is included in each coating run for destructive evaluation. This includes metallographic verification of coating thickness at designated locations on the fuel tube and continuous sigma layer present in the coating. A coating sample also underwent microprobe analysis to ensure proper chromium content in the coating.

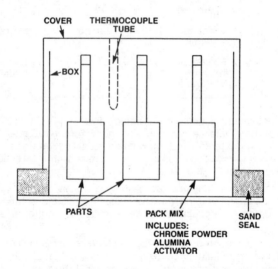

Figure 1 Schematic of Filled Container Used to Coat Parts in Pack Mix

Final release of the parts from the coating vendor is based on the satisfactory fulfillment of all of the above quality parameters. Numerous parts have been satisfactorily coated with this frozen process, and a high degree of confidence is placed on these controls to ensure successful chromizing.

COATING CHARACTERISTICS

Metallurgical

The chromized coating is an integral part of the Hastelloy B base material. The microstructure of the coating can be observed in Figure 2. The chromium gradient in this coating is shown in Figure 3; the percentage of chromium is high at the surface and gradually decreases to the low limits found in the base metal. The microstructure reveals a continuous layer on the surface of the coating. This layer is an intermetallic sigma phase, as seen in the Ni-Cr-Mo phase diagram[1] (Figure 4) and has been verified by X-ray diffraction. The remainder of the coating is a two-phase or nickel solid-solution.

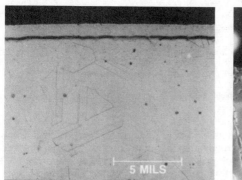 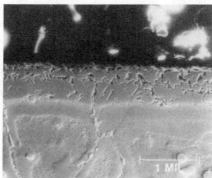

Figure 2 Typical Microstructure of Chromized Coated Hastelloy B

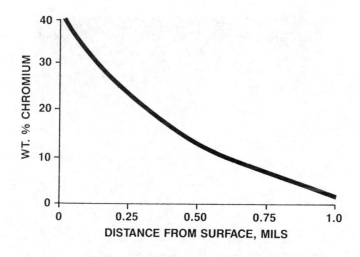

Figure 3 Composition Gradient in Coating

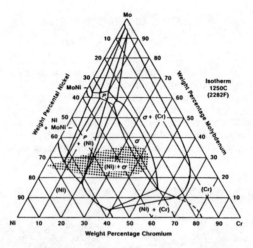

Shaded Area Represents Coating Compositional Range

Figure 4 Ni-Cr-Mo Phase Diagram

During the process development, metallographic work performed[2] disclosed that the characteristics of the coating were greatly affected by the coating thickness. At the upper limit of the chromized coating thickness (1.2-1.3 mils), the sigma layer was 0.2 mil thick and continuous, but thin coatings exhibited a discontinuous sigma layer (Figure 5). It was later noted that the coating thickness and therefore the existence of the sigma layer was directly related to the chromium percentage at the surface. As the sigma layer is felt to have a beneficial effect on corrosion behavior, a decision was made to assure the presence of a continuous sigma layer on all production coatings. This was verified during coating inspection.

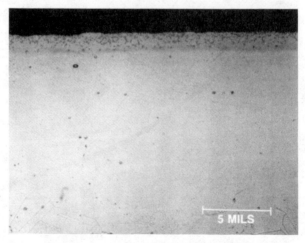

**Figure 5 Microstructure of a Thin Chromized Coating
With a Non-Continuous Sigma Layer**

Because the fuel tube is subject to assembly stresses, the ductility of the intermetallic σ layer was a concern. The specified stress relief cycle of 2165°F/8 min has been shown to increase the ductility of the coating. Bend testing showed that the coating could experience a strain of approximately 2 percent before cracking[3]. The increased ductility was believed to be caused by the inward diffusion of chromium, slightly lessening the thickness of the sigma layer. A 2 percent strain level was an acceptable value for assembly stress of the fuel tube.

A coated sample was subjected to a thermal cycle of 15 hours at 1400°F, followed by 81 hours at 1200°F to simulate the temperature variation observed by the fuel tube during its operational life. Upon completion of the test, the stability of the coating was excellent. No microstructural degradation, loss of adhesion, or breakdown in coating quality was observed.

The final metallurgical examination included determining the effect of the coating cycle on the base metal. No detrimental effects were noted. The coating cycle only increased the grain size from ASTM 4 to ASTM 3.5. As the coating temperature of 1940°F lies in the sensitization zone of the material for M_6C molybdenum carbides, some concern was raised; however, the stress relief cycle of 2165°F/8 min. was found to adequately solution all of the precipitated grain boundary carbides.

Environmental

The primary purpose of the chromized coating was to decrease the stress corrosion cracking susceptibility of the Hastelloy B base metal. Previous testing revealed Hastelloy B to experience SCC in ammonium hydroxide vapor, a decomposition product of the hydrazine fuel. SCC testing was conducted on chromized coated Hastelloy B in comparison to the uncoated base metal.

Stress corrosion testing was accomplished using C-ring samples per ASTM G38. The 0.75″ diameter by 0.75″ wide by 0.040″ thick samples were machined from Hastelloy B barstock. After coating, all of the samples were vacuum heat-treated at 2165°F/8 min. Loading was applied using a passivated 18-8 stainless steel bolt. Tightening the hex nut resulted in an increase in the tensile stresses on the outer surface of the specimen. An uncoated sample was strain-gaged to verify the loads placed on the surface of the samples.

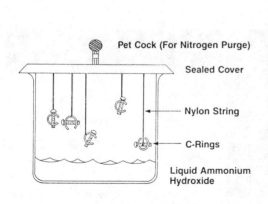

Pet Cock (For Nitrogen Purge)

Sealed Cover

Nylon String

C-Rings

Liquid Ammonium Hydroxide

Figure 6 SCC Bell Jar Test Setup

Testing was performed in ammonium hydroxide vapor. The C-ring samples were loaded to three stress levels: 10, 30, and 50 ksi. Three coated samples and two uncoated samples were tested at each stress level. One unstressed sample was also included as a control. The 17 samples were suspended above the ammonium hydroxide solution in a glass bell jar (Figure 6). Monthly inspection of the samples was performed. The ammonium hydroxide was changed after each inspection period and measured for ammonia concentration. Periodic gas samples were also taken to ensure a stable, consistent environment. On an average, the samples were exposed to vapor with approximately 45 percent ammonia.

The results of the stress corrosion tests can be found in Table C. The uncoated samples became severly discolored and highly pitted after very short exposure periods (less than one month). Stress corrosion cracks were observed on all but one of the samples after six months (Figure 7). The SCC cracks penetrated up to approximately 0.015″ depth. The chromized coated samples showed almost no effects of the ammonia environment (Figure 8). No cracking, discoloration, or extensive pitting was observed on the chromized samples (Figure 9). Testing on these samples was continued and has presently reached over 14 months of exposure. Still, no degradation of the coating has been observed. It has therefore been concluded that the chromized coating eliminates the stress corrosion susceptibility of Hastelloy B and improves general corrosion behavior.

Table C Result of SCC Testing of Chromized Coating and Uncoated Hastelloy B in Ammonium Hydroxide Vapor After 14 Months

Stress Level	Uncoated Hastelloy B	Chromized Coated Hastelloy B
0 ksi	No Cracks	No Cracks
10 ksi (20% F_{TY})	1 of 2 Samples Cracked After 6 Months	No Cracks
30 ksi (60% F_{TY})	Both Samples Cracked After 4 Months	No Cracks
50 ksi (100% F_{TY})	Both Samples Cracked After 4 Months	No Cracks
Overall	Susceptible to Stress Corrosion Cracking, Severe Pitting and Discoloration	No Stress Corrosion, Discoloration or Extensive Pitting

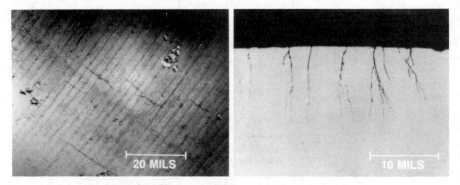

Figure 7 Stress Corrosion Cracking Observed on Uncoated Samples After 6 Months in Ammonia Hydroxide Vapor

176

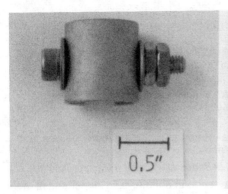

Figure 8 Chromized Coated Sample Tested in Ammonium Hydroxide Vapor

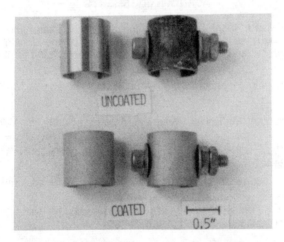

Figure 9 Chromized Coated and Uncoated Samples After 12 Months in Ammonia

A coated sample was run through a hyrazine compatibility test to make sure no degradation occurs in the sample due to exposure to the fuel. A 30-day test was conducted in a sealed Parr-bomb reactor. At the completion of the test, no significant rise in pressure was recorded, and no weight loss was noted. This indicated that no reaction took place between the coating and the hydrazine fuel. SCC testing was also conducted using coated and uncoated samples in hydrazine vapors. No stress-corrosion cracks were reported in any of the samples on coated or uncoated specimens; however, some pitting did occur on the uncoated samples.

All environmental testing was conducted on chromized coating that had a continuous sigma layer on the surface, as the sigma layer was believed to enhance corrosion resistance. Due to this fact, efforts were made to ensure a continuous sigma layer on all production pieces.

SUMMARY AND CONCLUSIONS

1. A chromized coating was chosen for a Hastelloy B fuel tube to reduce or eliminate the stress corrosion cracking tendency of the material.

2. A coating process was developed for the unique application on the fuel tube. Processing provisions and quality control checks were applied to ensure that a consistently uniform coating with adequate chromium would occur on all production samples.

3. Metallurgical characterization of the coating, including composition and identification of phases present, was carried out in detail.

4. The coating cycle was found to have no detrimental effects on the base metal.

5. The chromized coating was found to be compatible, and experienced no degradation in hydrazine propellant.

6. Chromized Hastelloy B was tested and found to be immune to stress corrosion cracking and general corrosion in ammonium hydroxide vapors within the test periods reported here.

ACKNOWLEDGEMENTS

The authors wish to thank Dallas C. Augustine of Sundstrand and Carl Gross of TCC for the roles they played in the development of the testing and coating development, and to Len Joesten and Owen Briles of Sundstrand for their laboratory assistance.

REFERENCES

1. Metals Handbook, 8th Edition, Volume 8, American Society for Metals (1973), p. 426.

2. Wentland, W.L., "Chromized Coating Development of Hastelloy B", Sundstrand Materials Report, MPR 27171, November 22, 1985.

3. Durako W.J., "Bend Limits on Chromized Hastelloy B", Sundstrand Memo, MLM 357-86, July 23, 1986.

4. Wentland, W.L., "Stress Corrosion Testing of Hastelloy B", Sundstrand Materials Report MPR 23930, January 21, 1985.

QUANTITATIVE PHASE ANALYSIS BY X-RAY DIFFRACTION

OF ZrO_2-8%Y_2O_3 SYSTEM

Marc J. Froning and N. Jayaraman

Department of Materials Science
and Engineering
University of Cincinnati
Cincinnati, OH 45221-0012

Abstract

A BASIC computer program for resolving overlapping diffraction peaks
has been developed. This program was successfully used to resolve the
$(004)_T$, $(400)_T$ and the $(400)_F$ peaks in several plasma sprayed ZrO_2-8%Y_2O_3
thermal barrier coating (TBC) systems. Results of structure factor analyses
and x-ray diffraction analyses of cubic, tetragonal and monoclinic ZrO_2
phases are presented.

179

I. Introduction

Quantitative phase analysis by X-ray diffraction in many complex multiphase systems depends on the ability to resolve overlapping diffraction peaks and on accurate structure factor calculations. Many methods have been proposed to resolve overlapping diffraction peaks with varying degrees of success (1-3). The major inadequacies of these earlier methods were that each peak was not considered to be made up of $K\alpha_1$-$K\alpha_2$ doublets and that a fixed shape (usually a gaussian) was assumed for the peaks. These assumptions could lead to erroneous quantitative analysis. In addition to the problems with resolving of peaks, the quantitative analysis is greatly influenced by an accurate determination of the individual phase structure factors, which in turn depends on the ion positions in the unit cell. For example, Miller and others (1) have used the structure factor data on ZrO_2-Y_2O_3 system published by Porter and Heuer (4). However, Evans and co-workers (5) pointed out that Porter and Heuer (4) incorrectly calculated the structure factors based on the ion positions given by McCullough and Trueblood (6). The ion positions recommended by Smith and Newkirk (7) are refinements of the ones suggested by McCullough and Trueblood and appear to work better. Evans (5) also pointed out that Porter and Heuer's work was based on MgO stabilized ZrO_2, rather than Y_2O_3 stabilized ZrO_2. The observed differences could then be attributed to the large differences in the atomic scattering factors of Y and Mg.

In this paper, we present an analytical method for resolving overlapping peaks and a structure factor analysis of ZrO_2 phases. The analytical method presented here treats the shape of the peaks as an independent variable and also considers each peak to be made up of a $K\alpha_1$-$K\alpha_2$ doublet. Both the peak resolving method and the structure factor calculation were experimentally verified.

II. Peak Resolving Method

The procedure is based on generating a composite curve from adjustable individual peaks and comparing the composite curve with the experimental curve. The parameters that are adjusted in the individual peaks are: (i) peak location of $K\alpha_1$ ($K\alpha_2$ location is dependent on the $K\alpha_1$ location), (ii) peak height of $K\alpha_1$ ($K\alpha_2$ height is assumed to be 0.45 times that of $K\alpha_1$ - this assumption is based on a large volume of data collected in single phases, both metal and ceramic), (iii) shape factor for $K\alpha_1$, (iv) shape factor for $K\alpha_2$, and (v) width of $K\alpha_1$ (the width of $K\alpha_2$ is assumed to be the same as that of $K\alpha_1$ - again, this assumption is based on a large volume of data collected in single phases, both metal and ceramic). In addition to these variables for individual peaks, the background is also treated as a variable. Therefore, the total number of peak parameters p, for a given set of data are (5 x m + 1) where m is the number of peaks fitted to the data. Then the intensity as a function of 2θ for the individual peak is determined using the equation

$$Y_i = f_{1i} H_i \exp \left\{ -\ln 2 \left[\frac{2(x - L\alpha_{1i})}{W_i} \right]^2 \right\}$$

$$+ (1 - f_{1i}) H_i / \{ 1 + [2(x - L\alpha_{1i})/W_i]^2 \}$$

$$+ f_{2i} 0.45 H_i \exp \left\{ -\ln 2 \left[\frac{2(x - L\alpha_{2i})}{W_i} \right]^2 \right\}$$

$$+ (1 - f_{2i}) 0.45 H_i / \{ 1 + [2(x - L\alpha_{2i})/W_i]^2 \} \qquad (1)$$

where

Y_i = intensity calculated at x (in angles of 2θ) for the ith peak.

f_{1i} and f_{2i} = shape factors for $K\alpha_1$ and $K\alpha_2$ (varies from 0 to 1) for the ith peak.

$L\alpha_{1i}$ and $L\alpha_{2i}$ = peak location of α_1 and α_2 in angle of 2θ for the ith peak.

H_i = peak height of $K\alpha_1$ for the ith peak.

W_i = peak width at half peak height for the ith peak.

The first two terms on the right-hand side of the above equation take the contribution due to $K\alpha_1$ into account and the latter two terms are for $K\alpha_2$. When the shape factor f = 1, the peak is gaussian shaped and when f = 0 it is lorentzian shaped; any fractional value for f will give a shape intermediate to these two extremes. The Y_i's for all the peaks at a given x are numerically added and the background intensity is added to this to get the total calculated intensity Y_c at x.

Subsequently, this Y_c is compared with the actual data collected from the diffractometer, and the errors between the calculated and the experimental intensity values are determined as

$$\Delta Y = Y - Y_c \qquad (2)$$

Now, in order to determine the corrections to the peak parameters, a Jacobian is used to establish linear equations relating small changes in parameters and ΔY according to

$$\Delta Y = J \times \Delta P \qquad (3)$$

where ΔP is the correction vector to the vector P. The Jacobian is formed by determining the partial derivatives of equation (1) with respect to the individual peak parameter. Therefore, the total number of elements in the Jacobian is p x p; for example, if we try to fit three peaks to a given set of data, the number of elements in the Jacobian will be 16 x 16 = 256.

This equation may be solved by simple linear algebra to yield ΔP's, found by inverting J and multiplying by ΔY to give

$$\Delta P = J^{-1} \times \Delta Y \qquad (4)$$

A new vector of parameters is formed by

$$P_{new} = P_{old} + \Delta P \qquad (5)$$

These new sets of adjusted peak parameters are used to calculate intensities. This whole process is repeated iteratively until the error mean square quantity

$$\xi^2 = \frac{1}{n-P} \sum_{1}^{n} \frac{(Y_c - Y_o)^2}{Y_o} \qquad (6)$$

is a minimum. Also, another criteria that is useful for determining the convergence is when the ΔP's are very small.

Computer programs were written in Applesoft Basic (on an Apple IIe with 128K RAM) and in GWBASIC (on an HP Vectra Model 25 with 640K RAM), and were run using Microsoft Basic Compilers. The flow chart shown in Figure 1 briefly explains the different steps involved in the computation. Many sets of experimental data were analyzed using the computer programs. Typically, a good set of data with three peaks converged in seven iterations and in about four minutes in the HP and in about 25 minutes in the Apple. A typical example of resolved peaks in the 400 region of ZrO_2 system is shown in Figure 2.

III. Structure Factor Analysis

Structure factors of diffraction peaks in tetragonal ZrO_2 and cubic ZrO_2 were determined assuming a CaF_2 structure for cubic ZrO_2 and the ionic positions recommended by Teufer (8) for the tetragonal ZrO_2. The atomic scattering factors for Zr^{4+} and O^{-2} ions were taken from references (9) and (10). In these analyses, the effect of the stabilizer Y_2O_3 were ignored. Using the structure factor calculations, relative intensities of the diffraction peaks (for $CuK\alpha$ radiation) were calculated. These theoretically calculated intensities were then compared to the experimental values.

IV. Experimental

The purpose of the experimental work was twofold: one to confirm the intensity values calculated in the previous section and, secondly, to verify the above analysis with known mixtures of monoclinic and tetragonal ZrO_2 and with known mixtures of cubic and tetragonal ZrO_2. For this purpose, three powders from another study (11) were used. These powders were (i) spray dried ZrO_2-8%Y_2O_3, which was found to have a mixture of monoclinic ZrO_2 and free Y_2O_3; (ii) fused, cast and crushed ZrO_2-8%Y_2O_3 which contained only tetragonal ZrO_2; and (iii) fused, cast and crushed ZrO_2-20%Y_2O_3, which contained only cubic ZrO_2. X-ray diffraction scans of all these powders were run and full scans are presented in Figures 3-5. In addition, slow step scans were performed in the regions of all individual

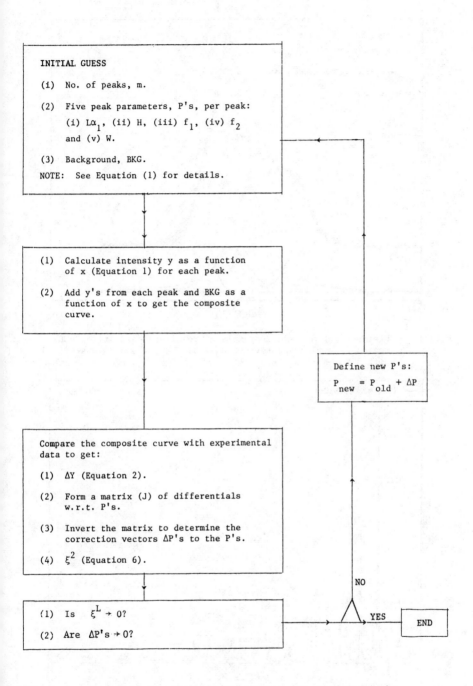

INITIAL GUESS

(1) No. of peaks, m.

(2) Five peak parameters, P's, per peak:
     (i) $L\alpha_1$, (ii) H, (iii) f_1, (iv) f_2
     and (v) W.

(3) Background, BKG.
NOTE: See Equation (1) for details.

---

(1) Calculate intensity y as a function
     of x (Equation 1) for each peak.

(2) Add y's from each peak and BKG as a
     function of x to get the composite
     curve.

Define new P's:

$P_{new} = P_{old} + \Delta P$

Compare the composite curve with experimental
data to get:

(1) ΔY (Equation 2).

(2) Form a matrix (J) of differentials
     w.r.t. P's.

(3) Invert the matrix to determine the
     correction vectors ΔP's to the P's.

(4) ξ^2 (Equation 6).

---

(1) Is $\xi^L \rightarrow 0$?

(2) Are ΔP's $\rightarrow 0$?

NO

YES

END

Figure 1. Flow chart for the computer program for resolving peaks.

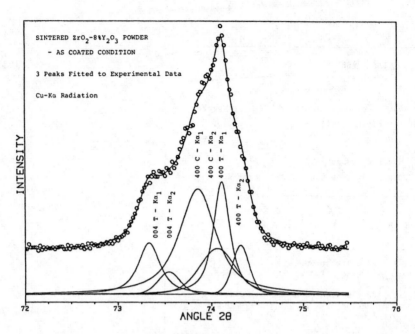

Figure 2. Plot showing a typical example of resolved peaks in the 400 region of a plasma sprayed ZrO_2-Y_2O_3 TBC. The data was fitted with 3 $K\alpha_1$-$K\alpha_2$ doublets.

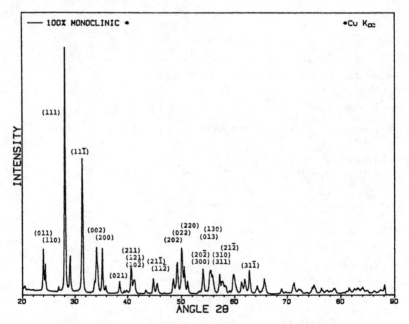

Figure 3. X-ray diffraction scan of monoclinic ZrO_2 powder using CuKα radiation.

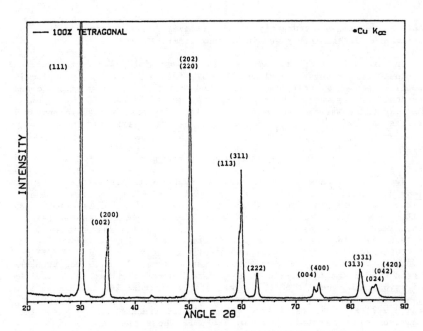

Figure 4. X-ray diffraction scan of partially stabilized tetragonal $ZrO_2-8\%Y_2O_3$ powder using CuKα radiation.

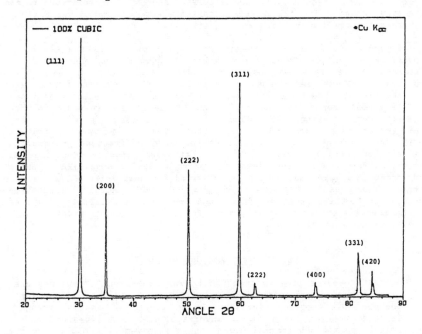

Figure 5. X-ray diffraction scan of fully stabilized cubic $ZrO_2-20\%Y_2O_3$ powder using CuKα radiation.

peaks to determine accurately their integrated intensities. The
experimental details are given in another paper (12).

In a second set of experiments, five powder mixtures of monoclinic and
tetragonal ZrO_2 powders, having 10%, 25%, 50%, 75% and 90% (by volume)
monoclinic contents, were prepared. Another set of mixtures of cubic and
tetragonal ZrO_2 powders, having 10%, 25%, 50%, 75% and 90% (by volume)
cubic contents, were prepared. Again, slow step scans of these powder
mixtures in the $(111)_{monoclinic}$ and $(400)_{cubic}$ were performed.

V. Results and Discussion

Tables I-III show the calculated and experimental integrated intensity
values for tetragonal, cubic and monoclinic ZrO_2-Y_2O_3 systems. The
experimental determination of integrated intensities of diffraction peaks
in the monoclinic ZrO_2 system were possible due to the peak resolving
method described in section II of this paper. A typical example of
resolved peaks for the monoclinic ZrO_2 system is shown in Figure 6. The
structure factors and the relative intensities of tetragonal ZrO_2-8%Y_2O_3
are shown in Table I. As seen in this table, there are some differences
between the values calculated in the present work and those calculated by
Teufer (8). As a result of this, the relative intensities calculated by us
differs from Teufer. However, the observed intensity values appear to
compare well with the calculated values of the present work. A similar
comparison for the cubic ZrO_2 in Table II shows that the maximum
discrepancy is for the (220) peak. All the other peaks seem to compare
well. As for the monoclinic ZrO_2, the results presented in Table III show
a very good comparison between the calculated and experimentally determined
intensity values. The calculated intensities were estimated using the
structure factors due to Smith and Newkirk (7).

Figure 7 shows the results of slow step scanning in the (111)
monoclinic region for the five monoclinic-tetragonal ZrO_2 phase powder
mixtures. These scans were used for volume fraction analyses and the
results of these analyses are presented in Figure 8. The plot of calculated
vs. actual volume fractions show a linear correlation and similar analyses
by Toraya (13), Porter and Heuer (4), and Evans (5) are presented in the
same plot for comparison. The attempts to do a similar analysis for the
cubic-tetragonal mixture were not successful. The difficulty was attributed
to (i) low peak-to-background ratio of the (400) diffraction peaks and
(ii) similar particle size of cubic and tetragonal powders resulting in
extreme differences in peak width; for example, a -325 mesh size in cubic
ZrO_2 produced a very sharp and narrow diffraction peak while a similar size
tetragonal produced very broad and shallow (400)(004) pairs of diffraction
peaks. Due to these problems, it was not easy to do a calibration similar
to the monoclinic-tetragonal mixture. Further studies are underway to
understand these problems.

VI. Summary and Conclusions

(1) A BASIC computer program that can be run on personal computers has
     been developed for resolving, accurately, overlapped diffraction
     peaks. This program determines, by an iterative process through
     least-squares fit and a matrix inversion, the peak location, peak
     height, peak width at half height, peak shape, and the background for
     the set of peaks fitted to the experimental data.

186

Table I

Structure Factor and Relative Integrated
Intensities of Diffraction Peaks in $ZrO_2-8\%Y_2O_3$ PSZ

(Fused and Crushed Powder)

Face Centered Tetragonal, a_0 = 5.135 Å C_0 = 5.185 Å

(copper Kα radiation)

$(hkl)_{BCT}$	$(hkl)_{FCT}$	F_C(Teufer)	F_C(NJ)	I_{cal}/I_{max} (Teufer)	I_{cal}/I_{max} (NJ)	I_{obs}/I_{max} (NJ)
101	111	30.7	32.6	100.0	100.0	100.0
002	002	21.9	23.9	9.4	9.9	9.6
110	200	18.3	20.3	12.8	14.0	12.7
112	022	29.7	34.1	30.2	35.6	39.1
200	220	31.7	36.7	17.1	20.4	22.7
103	113	22.2	26.4	11.5	14.6	12.5
211	311	21.8	26.3	21.8	28.2	26.2
202	222	17.7	21.1	6.5	8.3	7.8
004	004	18.5	23.6	1.3	1.8	2.1
220	400	21.9	29.4	3.5	5.6	5.2

Notes: (i) Calculated F_C (NJ) and I_{cal}/I_{max} (NJ) are based on Teufer's ion positions. Only Zr^{4+} and O^{2-} ions were considered for calculations.

(ii) The I_{cal}/I_{max} (Teufer) compares well with the calculated values reported in the JCPDS File 24-1165A.

Table II

Structure Factor and Relative Integrated
Intensities of Diffraction Peaks in $ZrO_2-20\%Y_2O_3$ FSZ

(Fused and Crushed Powder)

Cubic, a_0 = 5.141 Å   (copper Kα radiation)

hkl	F_C	I_{cal}/I_{max}	I_{obs}/I_{max}
111	32.6	100.0	100.0
200	20.3	21.2	23.6
220	36.7	61.4	49.3
311	26.3	42.5	39.2
222	19.0	6.7	6.5
400	29.4	8.5	12.0

Note: F_C and I_{cal}/I_{max} were calculated based on a CaF_2 structure. The atomic scattering factors of Zr^{4+} and O^{2-} were only considered.

Table III

Relative Integrated Intensities and D-Spacings of a Few Diffraction Peaks in ZrO_2-8%Y_2O_3

(Spray Dried Powder)

Monoclinic ZrO_2, $a_0 = 5.066$ $b_0 = 5.195$ $c_0 = 5.229$

(copper Kα radiation)

hkl	d,A	I_{cal}/I_{max}	I_{obs}/I_{max}
011	3.681	17	12.1
110	3.620	11	9.6
$\bar{1}$11	3.151	100	100.0
111	2.830	71	75.1
002	2.614	22	32.7
020	2.598	13	8.3
200	2.533	16	16.9
$\bar{1}$02	2.492	3	3.6
021	2.257	6	2.2
$\bar{2}$11	2.206	14	25.1
102	2.186	5	0.8
$\bar{1}$21	2.178	5	10.5
112	2.015	8	9.4
$\bar{2}$02	1.986	7	8.3
$\bar{2}$12	1.870	2	10.0
022	1.846	19	24.8
220	1.815	23	30.1
$\bar{1}$22	1.801	13	17.9
$\bar{2}$21	1.779	6	7.4

Note: In addition to these monoclinic ZrO_2 peaks, diffraction peaks from free Y_2O_3 were also found in the diffraction data.

188

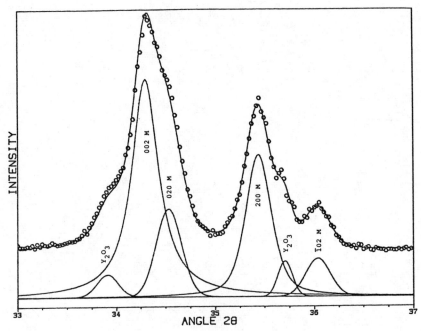

Figure 6. Plot showing a typical x-ray scan of monoclinic ZrO₂ with the individual peaks resolved.

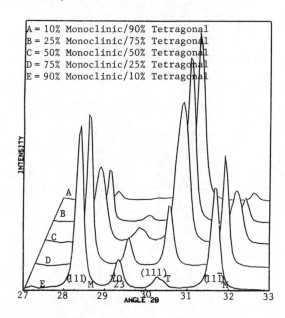

Figure 7. Plot showing x-ray scans of monoclinic-tetragonal ZrO₂ mixtures in the (111)_monoclinic region.

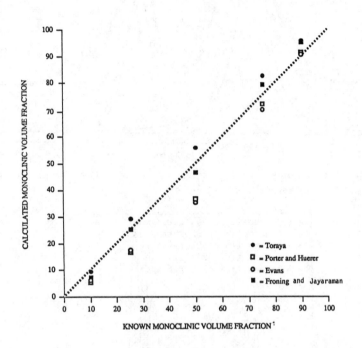

Figure 8.   Plot of calculated and actual volume fraction of monoclinic
phase in the monoclinic-tetragonal ZrO_2 mixtures scanned in
Figure 4.

(2) Structure factor analysis was carried out for tetragonal and cubic ZrO_2.

(3) Integrated intensities for diffraction peaks from monoclinic, tetragonal and cubic ZrO_2 phases were experimentally determined and compared with the predicted values from structure factor analysis.

(4) The above calculations were further verified by detailed experiments in mixtures of monoclinic-tetragonal ZrO_2.

References

1. R. A. Miller, J. L. Smialek and R. G. Garlick, in Advances in Ceramics - Volume 3, "Science and Technology of Zirconia," edited by A. H. Heuer and L. W. Hobbs, American Ceramic Society, 1981, pp. 241-253.

2. N. Ravishankar, C. C. Brendt and H. Herman, Ceram. Eng. Sci. Proc. 4(9-12), 1983, pp. 784-791.

3. R. D. Sisson, I. Sone and R. R. Biedermann, Thermal Barrier Coatings Workshop, NASA-Lewis Research Center, May 21-22, 1985, pp. 77-84.

4. D. L. Porter and A. H. Heuer, J. Am. Ceram. Soc. 62(5-6), 1979, pp. 298-305.

5. P. A. Evans, R. Stevens and J. G. P. Binner, Br. Ceram. Trans. J. 83, 1984, pp. 39-43.

6. J. D. McCullough and K. N. Trueblood, Acta Cryst. 12, 1959, pp. 507.

7. D. K. Smith and H. W. Newkirk, Acta Cryst. 18(6), 1965, pp. 983-991.

8. G. Teufer, Acta Cryst. 15, 1962, p. 1187.

9. International Table of X-ray Crystallography - Volume III, Kynoch Press, 1962, pp. 210-212.

10. B. D. Cullity, "Elements of X-ray Diffraction," Second Edition, Addison-Wesley, 1978.

11. M. J. Froning, N. Jayaraman, T. E. Mantkowski and D. V. Rigney in "Surface Modification and Coatings," edited by R. D. Sisson, ASM 1986, pp. 65-79.

12. M. J. Froning and N. Jayaraman - in this issue.

13. H. Toraya, M. Yoshimura and S. Somiya, Communications of the Amer. Ceram. Soc., June 1984, pp. C-119-C-121.

AUTOMATIC THERMAL IMPEDANCE SCANNING (ATIS) SYSTEM

FOR NDE OF COATINGS ON TURBINE ENGINE PARTS

D. R. Green and J. W. Voyles
Westinghouse Hanford Company
Richland, WA

J. H. Prati
Teledyne CAE
Toledo, OH

Abstract

An Automatic Thermal Impedance Scanning (ATIS) system has been developed for nondestructively testing thermal spray coating bonds in turbine engine parts. The ATIS NDE method is based on interpreting the infrared emitted from a test object's surface after injecting heat. Precision manipulators (robots) are used to load the parts into the system and to scan them under computer control. Signals from the scans are automatically computer interpreted to give probable locations of bond defects. The system was designed for practical day-to-day use in a production facility. This paper describes the system and some of the early results obtained. Thermal impedance methods can be used to detect nonbonds between metal parts and thermal spray coatings even when there are two layers such as a bond coat and a ceramic overcoat. Natural defects in production parts as well as fabricated defects in test specimens have been detected using these methods.

Acknowledgment: This project was sponsored by USAF/AFSC Aeronautical Systems Division, Wright Patterson AFB, OH 45433, under Teledyne CAE Subcontract #169278.

Introduction

Modern small gas turbines, as well as their larger aircraft gas turbine counterparts, place severe requirements upon materials in the engine. Typically, the materials operate at temperatures over 1900°F (1038°C) in the combustor and turbine sections. This temperature requirement, coupled with the needs for low cost, improved fuel efficiency, and durability has led to the extensive utilization of thermal spray coatings in gas turbines.

Thermal spray is a generic name for a family of coating processes such as wire spray, plasma spray, and detonation gun spray. Materials with special properties such as heat resistance can be applied to localized areas of a component to ensure survivability of the component and, in some cases, allow the use of lower cost materials.

Figure 1 shows a cross section of a small gas turbine and the typical locations where thermal spray coatings are utilized. In a small gas turbine, these can be classed into six generic groups:

1. Parts salvage
2. Clearance control (abradable seals)
3. Wear resistance
4. Thermal barrier
5. Corrosion control
6. Other specialized applications

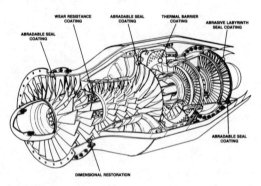

Figure 1 - Typical thermal spray applications in a gas turbine engine.

During the thermal spray process, particles of powder or wire are simultaneously heated to a plastic or molten state and accelerated to high velocities in a high temperature gas stream. The particles then impact the surface to be coated and bond to it by predominantly mechanical means. A good bond normally requires roughening the surface to be coated. Grit blasting is a popular method for producing this preroughening. Occasionally, the surface is grooved or threaded before grit blasting. This produces a rough, irregular interface between the coating and substrate. The outer surface of the coating is normally also rough (often comparable to medium or coarse grade sandpaper). Numerous inhomogeneities often exist in the volume of the coating, but these are normal and do not impair the performance in many applications.

Nondestructive testing of the bonds between coatings and their substrates is required to assure quality and reliable performance of engine parts. Thermal impedance scanning systems of the type described in this paper have been demonstrated capable of consistently detecting bond defects under thermal spray coatings. The primary purpose of the Automatic Thermal Impedance Scanning (ATIS) system is to provide completely automatic nondestructive examination (NDE) of thermal spray coating bonds in turbine engine parts. This system is being installed as part of an automated production facility at the Teledyne CAE plant in Gainesville, GA.

The ATIS NDE method is based on interpreting the infrared response of a test object's surface after injecting heat. An electronic infrared detector is used to scan the amount of infrared emitted. The time-temperature behavior of the test object which depends on its thermal surface impedance can then be interpreted to detect nonbonds under the coating. Successful application of this method to many coatings requires elimination of surface emissivity and reflection effects. Westinghouse Hanford Company (WHC) has developed methods for eliminating or minimizing the effects of emissivity and reflections in the ATIS results.

Past work has shown that thermal impedance methods can be used to detect nonbonds between metal parts and thermal spray coatings even when there are two layers such as a bond coat and a ceramic top coat. Natural defects in production parts, as well as fabricated defects in test specimens, have been detected using these methods. A system for marine shop applications delivered a year ago has been producing good results since delivery.

ATIS System Design

Theory of Operation and Design Approach

Thermal impedance is analogous to electrical impedance of conductors. A more proper term describing thermal impedance as used in the ATIS system is thermal surface impedance. It is the ratio of surface temperature to heat flow at the surface of the test object, just as electrical surface impedance is the ratio of voltage to current at the surface of a conductor. The surface impedance can be determined by injecting either a sinusoidal or transient heat flow at the surface of a test object and analyzing the resulting temperature versus time. Defects in a test object causes differences in the thermal impedance which manifest themselves as differences in the time-temperature behavior after heat injection. Thermal surface impedance has been analytically studied for the plane thermal wave case. The results of these studies were used to estimate the thermal surface impedance behavior with changing test object parameters. However, practical thermal impedance scanning rarely produces plane thermal waves. The actual dependence of thermal impedance on bonding was determined from calibration runs on prepared test specimens. Plane thermal wave theoretical results were used as a guide during calibration. This approach was similar to that successfully used for many years in eddy current nondestructive evaluation.

True sinusoidal thermal excitation of the test object is difficult since it requires cooling (as well as heating) to give negative (as well as positive) heat flow and has mainly been restricted to laboratory studies. Determination of the broad-band thermal impedance from transient temperature response requires analysis of a complete temperature transient from each point on the surface and requires a great deal of data as well as computer speed to provide a practical system. In a rapid scanning system, multiple infrared sensors and multiplexing of data into the computer would be required to produce enough data for computation of the thermal surface impedance in a large enough range of frequencies to provide separation of thickness, density, and bonding effects. The present ATIS system is aimed at detection of nonbonds only. Thickness measurement capability was accomplished in this system by adding a mechanical measuring probe to the ATIS sensor robot. Density of the coatings is not measured by the present system.

Bonding determinations are almost unaffected by moderate coating thickness and density variations when only low frequency components of the thermal transient are used in the ATIS analysis. Bonding influences the low frequencies considerably more than high frequencies for bonds located well under the coating's surface. The closer the nonbond is to the surface of the coating, the higher the predominant frequencies are that are influenced by the nonbond. Fortunately, the coatings of primary interest in the present work have their bond line far enough under the surface of the coating that detection of nonbonds requires only an approximation of the low frequency portion of the transient.

Analysis of the temperature transients for bonding effects can be further simplified by exploiting a prior knowledge of the general shape of the transient. By properly choosing the time at which the amplitude of the transient is sensed, a quantity that predominantly depends on the low frequency content and therefore on bonding can be determined. This approach is being used in the ATIS system being developed for TCAE.

A requirement that is often overlooked by workers in thermal NDE is that all points on the test object surface must have a thermal history (heat input versus time) that is known and is preferably simple and identical for all points. It is theoretically possible to determine the bond condition at each location if the thermal scan history is known, even if it is different from that of its neighbors. However, that approach is much more difficult and less practical than the one in which the heat input flux versus time, the total heating time, and the time elapsed before sensing the temperature transient are identical at each point.

A constant thermal history is produced in the present system by passing all points to be scanned through the same heat injection region so they all experience the same heat flux as a function of time. Constant delay time between heat injection and temperature sensing, as well as constant heat injection time, is achieved by rotating the part under the heat source and infrared sensor at a constant surface velocity and arc length between the source and sensor. This requires that both the speed of rotation and the angular displacement between the source and sensor be changed when the radius of the circular path being scanned is changed.

Temperature sensing in the ATIS system uses infrared methods that are independent of the emissivity of the part being scanned.

Computed Temperature Response

Although a number of materials have been scanned using methods similar to the ATIS approach, no experience existed in thermally scanning the specific materials of interest in the present work. Scanning parameters such as speed and heat input required for these materials were unknown. Theoretical computations were used to determine upper and lower limits of the scanning parameters for use in a preliminary design. Initially, no thermal properties were available for most of the coatings of interest. Thermal properties of materials believed similar to the ones of interest were used in the initial computations. Later computations were completed using thermal properties that were especially measured under the direction of Teledyne CAE by Purdue University[1] on the coatings of interest. System parameters were then redetermined and only minor revisions of the specifications were necessary even after completion of experimental engineering measurements on the required heat input, timing, accuracy, and detector characteristics.

Analytical equations for thermal wave propagation in simple cases contain simple terms that have physical significance such as wave velocity, attenuation, time constant, etc. However, analytical equations for transient temperatures in structures comprising several layers of materials in resistive contact are too complex to be useful, particularly when the thermal waves are not planar. The heat input versus time is complicated and is not precisely known. Finite difference methods can solve such problems with relative ease. For this reason, limiting cases for the time-temperature behavior were determined for use in the initial design of the ATIS system using finite difference methods, and the results were presented graphically for interpretation.

In order to complete a large number of computations in a short time, one-dimensional finite difference models for both lateral and through-transmission of heat were used. These resulted in upper and lower limits rather than exact answers, but they were accurate enough to establish limits for the design parameters. Care was taken to select theoretical models that would give conservative values for the limits. Figure 2 shows the two one-dimensional models (A and B) that were used to compute temperature transients that should result from two types of defects. Heat flow through the coating layers into the substrate in an infinite sheet with a uniform nonbond between the bond-coat and the substrate is modeled in Figure 2 (Model A). Heat flow laterally across the main coating layer over an infinitely long strip-shaped nonconducting bond defect is modeled in Figure 2 (Model B). These represent the limiting cases for heat flow through a large area defect (Figure 2, Model A) and a finite width, infinitely long strip-shaped laminar bond defect (Figure 2, Model B) with negligible conduction through the nonbond between the coating and a high conductivity substrate. A uniform heat flux of Q cal/sec-cm^2 is assumed to be injected into the surface for a limited time, and the surface temperature transient was computed for the time during heating and shortly thereafter. To simulate the heat-sink formed by the highly conductive substrate in the well bonded region, the coating temperatures at the edges of the defect were assumed to be fixed at T=0 in the computations. This gives a conservative value for the amplitude and duration of temperature transients expected over this type of defect. In all of the computations described in this report, the ambient temperature is assumed to be 0°, and all temperatures are in °C.

---

[1]R. E. Taylor and H. Groot, "Thermophysical Properties of Coatings," (A Report to Teledyne CAE), TPRL 477, School of Mechanical Engineering, Purdue University, West Lafayette, IN, September 1985.

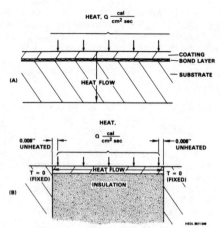

Figure 2 - One-dimensional heat transfer models. These models were used for finite difference computations. (A) is for through-transmission and (B) is for lateral heat flow through the coating over the defect.

These models may seem unrealistic. However, methods of this kind are often used by thermal NDE specialists to quickly determine required design limits since precise computations are often very time-consuming, and physical parameters such as shape of defects, as well as the heat input zone, are often not precisely known. Even if the defect size and heat input profile were known, the bond conductance varies from defect to defect; therefore, precise computations based on a representative defect would still be only an approximation.

The difference between the short term temperature transient over an infinite-area laminar defect and that over the center of a finite large-area laminar defect is small. Likewise, the difference in the transient along the center of an infinitely long strip-shaped defect and that over the center of a defect in which the length is only 4 to 10 times the width is small.

Although precise analytical computation of temperature transients in complex multilayer coatings with variable contact, defect geometry, and heat input profile is not practical, nonbonds can nevertheless be detected from differences they cause in the surface temperature transients during ATIS scanning. Finite difference computation of the approximate temperature transients, together with calibration and a knowledge of thermal wave behavior, can be used to select the proper scanning conditions and distinguish nonbonds from other variables such as moderate coating thickness and density variations.

Sets of computations representative of three gas turbine components were completed. The first was for a centrifugal compressor shroud which has an aluminum-6% silicon top coat and a nickel-20% aluminum bond coat on an A286 stainless steel substrate. The second set represented a compressor stator assembly. This assembly has a top coat comprising 60% mixture of Al-12% Si alloy with 40% polyester (AlSi-polyester) and a bond coat of 93% nickel-4% aluminum (NiAl) on a cast substrate of C355 aluminum. The final set represents an axial compressor rotor which has the same coating as the

compressor stator assembly but is applied to a Ti-6% Al-4% V titanium
substrate.

Initial Experimental Engineering Measurements

Experimental engineering measurements were completed to confirm the
theoretically predicted design of the ATIS system prior to construction of
some of the more critical parts. These measurements were performed using a
laboratory experimental thermal impedance scanning system to scan cylindrical test specimens which represented the three cases mentioned above and
contained fabricated bond defects. The approximate effective values for the
required heat input and infrared signal sampling, scanning speeds for
different types of coatings, and the emissivity compensation factor were
confirmed and refined by the measurements. In addition, they provided
information on the sensor design and accuracy requirements. Although no
major changes in design parameters were found to be necessary, several small
design adjustments were made to improve the performance of the prototype.
Such preliminary measurements are very helpful when designing a complex new
system.

An assortment of cylindrical thermal spray coated specimens used to
obtain the experimental results, as well as one of the turbine parts used
later to test the ATIS system, are shown in Figure 3. Coatings on the
cylinders represent the range of thermal properties of coatings on turbine
engine parts that must be nondestructively examined during the initial
application of the ATIS system. All of the cylinders contained artificially
produced nonbonds representative of real bond defects that can occur between
the coatings and their substrates. Scan results on most of the defects, as
well as destructive examination results from one of them, indicate that the
coatings over the defects are weakly bonded rather than totally nonbonded
and separated from the substrate. The defects were produced by applying a
thin coat of "DYKEM" blue layout fluid in spots ranging from 1/16" (1.6 mm)
to 1/2" (12.7 mm) in diameter on the substrate after grit blasting and
before the thermal spray coatings were applied.

Figure 3 - Cylindrical test specimens and centrifugal
compressor shroud containing fabricated bond defects.

Figure 4 shows the laboratory experimental scanning system during a
scan on one of the cylindrical test specimens. A lathe was used to rotate
the test specimen, and a cross slide driven by a stepping motor was used to
translate the heat source and infrared sensor during a scan. In this way,
the heat source and sensor traversed the test specimen in a helical path. A

map of the thermal impedance dependent quantity's amplitude is plotted as a function of distance along each circumferential scan line around the helix, and each map line is displaced by an increment that represents the displacement along the length of the test specimen. In this way a pseudo two-dimensional isometric map is produced in which angular displacement around the test specimen is represented from left to right in the map, and both longitudinal displacement and thermal impedance amplitude are represented in the vertical direction.

Figure 4 - Laboratory experimental thermal impedance scanning system.

The experimental heat source, seen above the cylinder in Figure 4, incorporated an oxy-acetylene flame. Computations showed that at high scanning speeds, the flame would heat the surface of aluminum alloy coated test specimens to only about 200°C, while temperatures deep within the coating would remain much cooler than this. A high intensity heat source is essential for successfully scanning at high speeds to detect defects in high thermal conductivity coatings. The scanning time for each of two passes required on an entire 3" (7.6 cm) diameter x 5" (12.7 cm) long coated area was about 7 seconds for aluminum alloy coatings up to 0.07" (1.8 mm) thick.

The infrared sensor, seen to the left of the cylindrical test specimen in Figure 4, incorporated an infrared photovoltaic detector. This type of detector requires liquid nitrogen cooling, but it has the required frequency response and detectivity for high speed scanning.

Considerable processing of the data is required to eliminate unwanted variables, obtain emissivity independence, and control the experimental system during the high speed scans. A Hewlett Packard 9000 Model 320 was used in this function since it has the required speed and peripherals.

Results from a scan using the laboratory experimental system on one of the cylindrical test specimens are shown in Figure 5. The coating on this specimen is 0.070" (1.8 mm) thick wire sprayed aluminum-6% silicon over a 0.003" (.08 mm) thick nickel-aluminide bond coat, and the substrate is stainless steel. Defect indications seen in the top half of the figure are from intentionally introduced 1/8", 1/4", 3/8", and 1/2" (3.2, 6.4, 9.5, and 12.7 mm) diameter circular defects and from a single teardrop shaped defect approximately 1" (25.4 mm) long x 1/4" (6.4 mm) wide at its widest point. The substrate thickness in this region, i.e., in the top half of the thermal impedance map shown in Figure 5, is 1/16" (1.6 mm). The substrate in the bottom half of the map is 1/8" (3.2 mm) thick, but the defects are otherwise identical to those indicated in the top half.

Figure 5 - Emissivity-independent infrared thermal impedance
scan map showing fabricated bond defects.

A side-to-side variation in the scan lines across the map can be seen
as a slow, single undulation in some of the scan lines, particularly at the
bottom of the map. These are due to initial temperature gradients that
built up from one side of the cylinder to the other due to convection
currents while it was stationary waiting for the scan to begin. This was a
consequence of the semimanually operated laboratory experimental system and
should not be a problem in the final robotic system since the test specimen
is handled much more quickly.

Figure 5 shows the power of the emissivity-independent infrared thermal
surface impedance scanning method. Although the background noise in this
thermal impedance map is rather high, defect indications down to 1/4" (6.4
mm) in diameter can easily be distinguished. Figure 6 shows the scan map
from an ordinary transient infrared thermogram on this sample made using the
same scanning conditions and heat input intensity. Note that not even the
largest defect indications can be seen above the emissivity noise in the
ordinary thermogram.

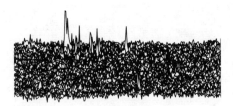

Figure 6 - Ordinary transient infrared thermogram on the same
test specimen as in Figure 5 using the same scanning speed
and heat input.

Natural "berry" defects in the plasma spray coating on the cylinder
shown second from the right in Figure 3 caused the large peak indications
seen in Figure 7. The coating on this test specimen is a plasma sprayed
50%-50% mixture of aluminum-12% silicon powder and pure aluminum powder over
a 0.004" (0.1 mm) thick nickel aluminide bond coat. The substrate is 1/16"
(1.6 mm) thick stainless in the top half of the map and 1/8" (3.2 mm) thick
stainless in the bottom half. "Berry" defects are visible as seen in
Figure 3. However since the ATIS system must be entirely automatic, it must
be capable of distinguishing this type of defect as well as bond defects
unaided by human intervention.

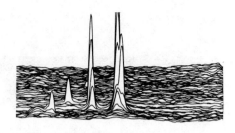

Figure 7 - Emissivity-independent thermal impedance scan map showing
peaks resulting from natural "berry" defects in the coating.

Figure 8 is a thermal impedance scan map of the mid-diameter region in
the test specimen shown second from the left in Figure 3. This specimen has
holes running longitudinally the full length under the surface of the large
diameter section of the substrate which, in this case, is aluminum. The
coating is a 0.073" (1.8 mm) thick layer of aluminum - 40% polyester powder
over a 0.005" (0.13 mm) thick nickel aluminide bond coat. In this specimen,
as in the others, a pattern of fabricated bond defects has been produced.
The thermal impedance map shows a cyclic variation resulting from the
subsurface holes in the substrate as well as the bond defects. Several
methods can be used to eliminate the effect of the substrate features. The
preferred method obtains its output from the part of the thermal impedance
signal that depends primarily on the higher frequency part of the thermal
wave response at the specimen surface. Due to limitations in the initial
laboratory experimental setup, it would have been time-consuming to set up
this method, so it was deferred until the fully automatic system could be
used.

Figure 8 - Emissivity-independent thermal impedance scan map showing
fabricated bond defects and subsurface holes through the substrate
(see text).

Figure 9 shows scan results indicating that the "DYKEM"-produced
artificial defects are weakly bonded rather than completely nonbonded with a
laminar crack at the interface. The thermal resistance of even a thin
laminar crack is so much higher than that of the coating material, it almost
completely blocks the short term heat flow from the coating into the
substrate. Therefore, there is little difference between the short term
thermal surface impedance of the coating over a flat-bottom hole and that
over a delamination covering the same area and the same depth under the
surface as the bottom of the hole. Thus, a flat-bottom hole should cause a
thermal impedance scan signal variation that reasonably approximates that
over a laminar crack.

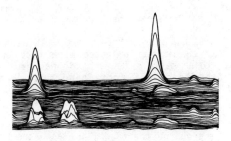

Figure 9 - Emissivity-independent thermal impedance scan map showing
the comparison between total nonbonds (largest two peaks) and
"DYKEM" produced bond defects (see text).

In the top part of the map in Figure 9 are two large peaks. These
correspond to holes milled through the substrate up to its interface with
the coating. The largest hole is approximately 5/16" x 3/8" (7.8 mm x 9.4
mm) and the smaller one is 1/4" x 5/16" (6.3 mm x 7.8 mm). This was done
early in the ATIS system development to obtain signals large enough for
initial alignment of the experimental system. In addition, as part of an
initial test, the test specimen was painted black to increase the signal-to-
noise ratio. Two other peaks that are very uneven are seen at the lower
left side of the figure. These are due to regions where the coating was
removed during destructive analysis to determine the degree of bonding. One
of these was located over the original 1/2" (12.7 mm) diameter "DYKEM" bond
defect in the lower part of the map. All of the other peaks are due to
"DYKEM"-produced bond defects. Note that these thermal impedance peaks are
much lower in amplitude than the ones over the flat-bottom holes.

The destructive analysis performed on this specimen comprised cutting
through the coating around the 1/2" diameter "DYKEM" defect and determining
how difficult it was to pry the coating off. In this test, it was found
that the coating was weakly bonded over the "DYKEM", requiring a moderate
force to pry it away from the substrate. An identical control test in an
area where the scan results indicated good bonding showed that the coating
was very well bonded and could not be pried off in a single piece.

Automatic Thermal Impedance Scanning (ATIS) System

The ATIS system was designed for automatically scanning turbine engine
parts to detect bond defects between the coating and substrate. Signal
interpretation as well as the scan itself are carried out under computer
control without human intervention. Test objects are handed to the ATIS
system by a transfer robot in an automated thermal spray production line
under command of a central process control (CPC) computer. The command to
begin the scan is sent from the CPC computer to the ATIS computer, and an
accept/reject signal, as well as an "OK to transfer" and other part transfer
signals, are sent to the CPC computer from the ATIS computer.

Scans are conducted at high speeds by the system, limited mainly by the
propagation time for the thermal wave to pass through the coating(s).
Typically, scans on the cylindrical test specimens having up to 0.070"
(1.75 mm) thick aluminum coatings require about 4 to 7 seconds for each of
two required passes. Total scanning time, including handling and position-
ing of the parts, is expected to be approximately 1-1/2 minutes from the

time the robot picks up the part until it is ready to release it. Less conductive coatings, such as aluminum-polyester mixes, will require perhaps 30 seconds longer due to the slower heat propagation through them.

At the time this paper was written, the ATIS system was still being programmed for some of the more complicated moves necessary to scan coatings on complex turbine engine parts. However, it has been tested in scans on the cylindrical test specimens containing fabricated defects and was demonstrated capable of detecting bond defects and mechanically measuring coating thickness.

Figure 10 shows an overall view of the ATIS system. The robotic scanning subsystem is shown on the right, a plasma arc heat source is shown in the background, and the electronic control and data analysis subsystems are on the left. A Hewlett Packard Series 9000 Model 320 computer is contained in the data analysis section. The plasma arc heat source was built by Plasmadyne, Inc., and the robotic subsystem was custom built by Anorad Corp. to WHC's specifications. A feature of Anorad robots essential to the success of the ATIS system is their ability to synchronize all of their movements to the rotational position of a spindle.

Figure 10 - Automatic Thermal Impedance Scanning (ATIS) System.

A detailed view of the robot subsystem is shown in Figure 11. This system contains two robots; one manipulates the infrared sensor, and the other manipulates the heat source during a scan. A specimen cylinder (white object) is seen mounted on the spindle assembly just above the center of the figure. The sensor robot is on the left (see label A) and the heat source robot is on the right in the figure (see label Q). A closeup of the spindle, sensor (on left inside of part), and source (on bottom inside of part) is shown in Figure 12. Figure 13 shows the spindle assembly and sensor with the infrared sensor rotated out of the way and the mechanical thickness probe rotated into position, ready for a measurement. The heat source is retracted out of the way during this operation. The system is capable of spindle speeds from 0 to 800 rpm and linear heat source and sensor motions at velocities from 0 to 10" per second.

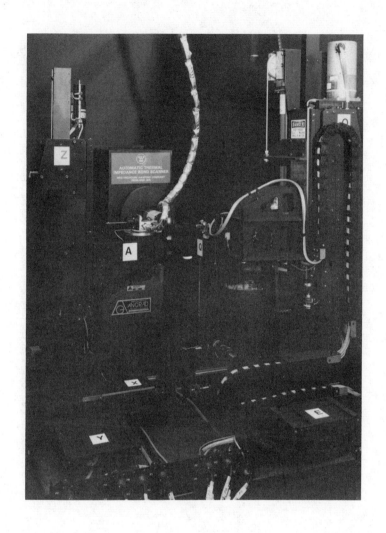

Figure 11 - Robots and spindle/turntable in the ATIS system.

Figure 12 - Spindle/turntable, infrared sensor, and heat source in position on a turbine axial compressor stator.

Figure 13 - Spindle/turntable and ATIS system's thickness measuring probe in position ready to measure coating thickness.

Figure 14 shows the control (left side) and the signal analysis (right side) subsystems. Numerous software routines are being written for the system to provide not only the various required complex robot movements under command of the data analysis and acquisition computer, but to provide several different types of data analyses and presentations for the convenience of the operator during setup and diagnostic work. These are in addition to the software already completed for the thermal impedance scan analysis and output map presentation.

Figure 14 - Electronic control, data analysis, and display subsystems.

ATIS Data Display

Although data interpretation in the ATIS system is normally performed by the system's computer, graphic display capability has been provided for engineering examination and process diagnostic work as well as setting the system up for new types of samples. Data from ATIS scans are archived on 3-1/2" micro disk (710 Kbytes) and temporarily stored on the system's 24 Megabyte hard disk. The graphic display software can be used to review current data from the hard disk or archived data from a micro disk. Hard copies can be produced using the system's printer in its graphic mode or the data can be quickly viewed in an identical display on the CRT at the operator terminal.

Figure 15 shows a graphic display of the map from an ATIS scan on a cylindrical test specimen having a wire sprayed 0.035" thick aluminum-6% silicon coating over a 0.003" thick nickel aluminide bond coat. The substrate cylinder wall for the region represented by the top half of the map is stainless steel 1/16" (1.6 mm) thick, and it is 1/8" thick (3.1 mm) for the bottom half. At the time this scan was performed, no feature for eliminating substrate thickness effects had yet been incorporated into the ATIS system software. The change in thickness caused crowding together of several scan lines in the middle of the map. This is seen as a dark band through the middle of the map. The dark band seen at the bottom of the figure was caused by end effects. Provision for minimizing thickness and end effects is being incorporated in the final system software.

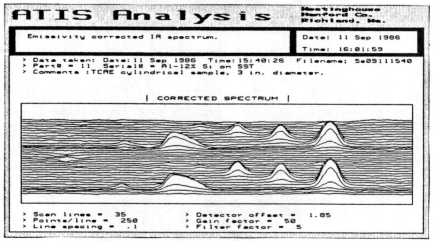

Figure 15 - Data display from an ATIS scan.

Bond defects can easily be distinguished from other features in the ATIS map on this particular test specimen, even though not all of the powerful analysis capabilities had yet been included in the data analysis software at the time this test was conducted. From right to left in Figure 15, each of the two sets of defect indications in the map are from circular "DYKEM" defects 1/2", 3/8", 1/4", and 1/8" (13 mm, 9.4 mm, 6.3 mm, and 3.1 mm) in diameter, a teardrop shaped defect approximately 1/2" long x 1/4" wide at its widest point, and another defect approximately 1/8" in diameter just beyond the teardrop.

Conclusion

Research and development on thermal surface impedance NDE technology has been underway in bits and pieces under various government and industrial contracts during the past 25 years. A large enough reservoir of technology has been developed that it has now become a field of powerful methods. This has made it easier for engineers skilled in this field to select NDE applications in which this technology has advantages. Nondestructive testing of coatings, especially coatings that are inhomogeneous and have rough surfaces and interfaces, is such an application. The speed, repeatability, and ability to ignore surface roughness, as well as its adaptability to automatic computerized data analysis, make the thermal impedance technology ideal for robotic production NDE applications. Added to the apparent ability in at least some cases to detect weak bonds as well as total nonbonds, these advantages have resulted in the present program to develop the ATIS system. The integration and demonstration of technology that have resulted from this program should be of definite benefit to industry.

Subject Index

Author Index